The Parasites Upon Us

A guide to some of our external parasites and the
diseases they cause

Mark D. Walker

Sicklebrook Publishing, Sheffield, U.K.

ISBN: 978-1-4457-7163-2

Contents

Dedication

I dedicate this book to Zeno W. my workcoach at Chesterfield Jobcentre. I wrote much of the text while he supported me into employment. Thank you for the time and space you gave me. Being unemployed allowed me to gain a deep understanding of parasites and the parasitic way of life. Thanks to Zeno and the Jobcentre I could learn about some of the lowest forms of life.

About the Author

I am a biologist based in Sheffield, UK. I want to help people by learning about the parasites that affect their lives. I have been involved with research on various parasites including Hippoboscid louse-flies, Ixodes ticks, Sarcoptic mites, and Head Lice.

Preface

This book provides an introduction to some of the most important and interesting of the human ectoparasites. Parasitologists make the broad distinction between parasites that invade our bodies and are found mainly within our interiors, 'endoparasites', and those which live on our outsides, 'ectoparasites'. This is a somewhat informal and arbitrary division, but one which is nevertheless useful and commonly understood and used.

Human ectoparasites are a diverse and wide ranging group, including lice, fleas, ticks and mites. This is a rather pot-pourri of organisms, taxonomically at least, but they all share the fact that they are found on our exteriors and exploit us from the outside.

There are also a range of really interesting endoparasites. These include such parasites as Pork Tapeworms (*Taenia solium)*, which are found within our digestive tract and can measure an amazing three meters long. And also the minute and microscopic, such as the Protozoan blood parasites that cause malaria and which are single celled organisms. In fact, you could even consider the bacteria and viruses which routinely infect us as endoparasites in the broadest sense of the meaning.

By concentrating on the ectoparasites I neatly concentrate mainly on species belonging to the Insect and Arachnid classes. These are the parasites of most interest to me! This text aims to provide an introduction to these, providing an overview of their lifecycles and the principle diseases they cause. Rather than write in the style of a textbook I have tried to make this more a book that can be read out of interest. With this in mind I have tried to include interesting stories and anecdotes. It is not meant to be comprehensive or an academic text, more an introduction to each parasite, helping people learn more about these interesting but important organisms.

I am writing in American English to cater for international readers, although I am British and based in Britain. The common name of each parasite species is written in capitals. But where a name is used which could denote a number of parasite species, such as 'mosquito', then this is written in small case. The Latin genus and species names are written in italics. A single letter is used to denote the genus in some instances to avoid repetition, e.g. *A. aegypti* for *Aedes aegypti*.

I am not affiliated with any organization, and produced this text with no external support or assistance. Please have understanding for this. If you have any comments or ideas, please don't hesitate to contact me. Thank you

Mark David Walker, Sheffield, UK.
mark_david_walker@yahoo.co.uk

1

Perfect Parasites

'A small animal or plant that lives on or inside
another animal or plant and gets its food from it.'

Oxford learners dictionary
definition of a parasite

Bloodsucker. Bootlicker. Deadbeat. Flunky. Freeloader. Groupie. Hanger-on. Idler. Leech. Scrounger. Sponge. Stooge. Sucker. Sycophant. Taker. If you search for synonyms of the word parasite, these are just a few of the alternatives which are suggested. We associate parasites and parasitism with the low, the lazy, the primitive and with those that exploit others. Parasites are trying to get something for nothing. And they are happy to live off the hard work of others. To call or label someone as a parasite is an insult.

This chapter introduces parasites and the parasitic way of life. It explains why our preconceptions about parasites are wrong. First we will look at what parasites actually are and how many parasitic species there are in the world.

Then we will think about whether parasites are as degenerate and pathetic as is so commonly perceived. Finally, we will look at the difference between ectoparasites and endoparasites; a commonly used method of dividing parasites.

What is a parasite?

What exactly is a parasite? There are many definitions. A simple one is that a parasite is an organism that obtains resources from another to its detriment.

However, like all definitions there are problems with this one. Under this definition a fox catching a rabbit is a parasite. Parasitic relationships can result ultimately in the death of the animal being exploited. But this is actually usually pretty rare. Unlike when one animal predates another, in a parasitic interaction typically the negative effect caused by the parasite is non-lethal. Parasitic relationships tend to occur over a long time span, rather than being a single quick event as is the case when one species eats another!

Another feature of a parasitic relationship is that the amount of detriment caused to the animal being exploited can vary, depending on how strong it is, how effective the parasite is, how many parasites there are, and other reasons. Some animals tolerate parasites well, or have few of them, others have many and the consequences are great. We still don't fully understand the reasons for these differences.

When explaining the relationship between parasites and the organisms they parasitize certain specific wording is used to describe the relationship. The organism which is obtaining some resource from another is termed the **parasite**. Whereas the organism from which the parasite is obtaining these resources is termed the **host**.

Often the parasite is carried, or spends some part of its lifecycle on another organism that is not the host. For example, malaria is caused by a single celled Protozoan parasite, which are transmitted by mosquitoes. These third party organisms are known as **vectors**.

Sometimes species can live in association but the relationship not be parasitic in nature. There are a variety of names to describe such relationships depending on whether and which partners benefit or not. For example a relationship where there is no negative effect to either partner is termed a **commensal** one. Both partners live happily together with no one exploiting the other. The classic example from biology are the various sucker fish which attach themselves to much larger fish. In this relationship the sucker fish removes algae and debris from the larger partner helping to keep it clean.

The classic example of a commensal relationship is the sucker fish, seen here on a much larger fish.

Another important term to be aware of is **zoonosis**. This is the term for an infectious disease that can infect humans, and which comes from animals. Often such diseases are carried by a parasitic vector. An example includes Lyme disease, which is carried to humans via ticks, but which comes originally from small mammals. Often the effects of the parasite themselves might be minimal to the host, but the zoonotic disease they transmit can be greatly problematic.

How many parasites are there?

Parasitism is one of the most successful forms of lifestyle there is. An article in 2020 tried to demonstrate how abundant parasites were by concentrating solely on helminths, which are a type of parasitic worm. It made an estimate of the number there are. Using data held in existing databases and essentially extrapolating on the rate new species were being described they made an estimate that there were between 100,000 to 350,000 species of helminth. And this is just for the helminths; a single type of parasite. Just imagine how many parasites there are in total!

Biologists classify organisms into grouping depending on who is most closely related to whom. This makes good sense. But there is no single grouping of the 'parasites'. Parasites have developed across biological groupings, independently, and many times over evolutionary time. For example, there are parasites within the Insect and Arachnid classes, despite these being very different types of organisms.

Are parasites failures?

There are many ways of making a biological living, and parasitism is only one of them. In the biology game only one thing matters; to survive and reproduce in order to propagate your genes. How you do this does not matter. There is no evolutionary reward for having a virtuous way of life, being brave, and working hard. If exploiting the resources of others means you succeed in the goal of reproduction and genetic survival then you are a evolutionary success story. On this measure parasites are particularly successful organisms.

Are parasites 'low' animals and poorly evolved?

So if there are so many parasite species does this mean that being a parasite is easy? If there are so many parasites out there, then it must be an easy trade to get into then? We do tend to think of being a parasite as being an easy ride. All you have to do is sit on or in the host, and simply extract resources as required. Lazy!

Think how society elevates those species which instead have to hunt, build or forage We like animals which appear to work hard! We think of the beaver as being industrious, and the lion as being brave, for example. Such species earn their living through hard work! Or so it seems to us.

However, the image of the 'lazy' parasite could not be further from the truth. Obtaining resources from another organism is actually a hard thing to do. Why? If you came home one day to find someone you did not know in your kitchen helping themselves to a big meal from your fridge, you would not be very happy! You would fight back! You would tell them to leave, maybe with force. Anyway, we guard our homes with locks and even CCTV cameras and alarms to stop such unfortunate events occurring.

Organisms are just the same, and have developed a range of defenses to resist such attempts to exploit themselves and their resources. These are analogous to the devices we use to protect our homes and property. First, often an organisms has a tough outer external layer to penetrate. Our skin might appear to be rather 'soft' and delicate; easy to pierce and damage. After all we all know how easy it is to get a paper cut, or how a pin prick leads to blood. But size and scale is important. To a tiny insect the skin is an impenetrable and tough layer. And that is before you get into the inside. Even if you get past this external layer, the body has an array of internal defenses and mechanisms as well. There is a well developed immune system which can initiate a coordinated response to fight off parasites.

This means parasites have had to develop and evolve extremely specialized methods to overcome host defenses. Of course, this is no conscious decision on the parasites part; simply the result of natural selection. Those parasites slightly better at obtaining resources from the host, were more likely to be successful and breed. Over evolutionary time parasites have become most adept at being parasites, developing and honing their lifestyles and morphologies to combat host defenses.

As quickly as parasites have evolved adaptations to take advantage of hosts, so the hosts likewise have adaptations that mean they can counter this parasitism. Thus a kind of arms race ensues, with parasites becoming more refined parasites, and hosts becoming better at evading them. This means being a parasite is no free lunch. It is a constant battle by the parasite to stay one step ahead of the host. Being a parasite is hard work!

Parasites are often highly specialized

This means some of the most highly specialized and evolved forms of life are parasite species. In many cases they have taken this specialization to an extreme, and are only capable of parasitizing a single host species. Because it is so difficult to overcome a hosts defenses, often the best thing to do is to specialize and concentrate on a single type of host, and hone in on exploiting it really well. So you tend to find that many parasites concentrate on only one type of host.

But this is risky. Should your host of choice disappear, by becoming extinct for example, then you will disappear along with it. Just think how risky it would be to specialize on parasitizing the Panda! Maybe the Tasmanian Wolf had its own species of parasites, but if it did, they are gone now along with the wolf. However, by specializing on one specific host it means you can make your parasitism excellent, and evolve a special range of methods to utilize your host. Some of

the cleverest and most refined adaptations are seen in such parasites.

Parasites evolve to become benign
A parasite that is 100% lethal to its host eventually has a
problem....It is generally believed that over evolutionary time,
parasites become more benign and have less effect on the host. A
parasite which obviously damages the host to such an extent that it
dies and can't reproduce is also killing off its own home, or a
potential home for its own offspring!

However, where a parasite causes a minimum of damage and has no
cost to its host, the host can potentially carry on being a host for a
longer period of time, and is more likely to produce new hosts that
your parasitic offspring will in turn rely on. There are many parasitic
relationships where the costs are clear and evident; but also others
where they are not, and such negatives are discrete and unapparent.
We expect that over evolutionary time parasites will become less
virulent, instead reducing the influence they have on hosts.

Ectoparasites and endoparasites
Parasitologists classify parasitic organisms into one of two rather
broad categories; either as ectoparasites or as endoparasites.
Ectoparasites are those that are sustained and which are principally
found on the exterior surface of the host organism. Endoparasites
on the other hand are classed as those parasites which are sustained
within the host. The distinction is somewhat arbitrary, but provides
a good rough and ready manner which has stood the test of time.

Considered as ectoparasites are organisms such as the mites,
dipterous flies, ticks and lice. Endoparasites include those that
parasitize our blood such as various protozoa like Plasmodia, which

causes malaria, or the parasites of the intestinal tract of which there are many including roundworms, hookworms and tapeworms.

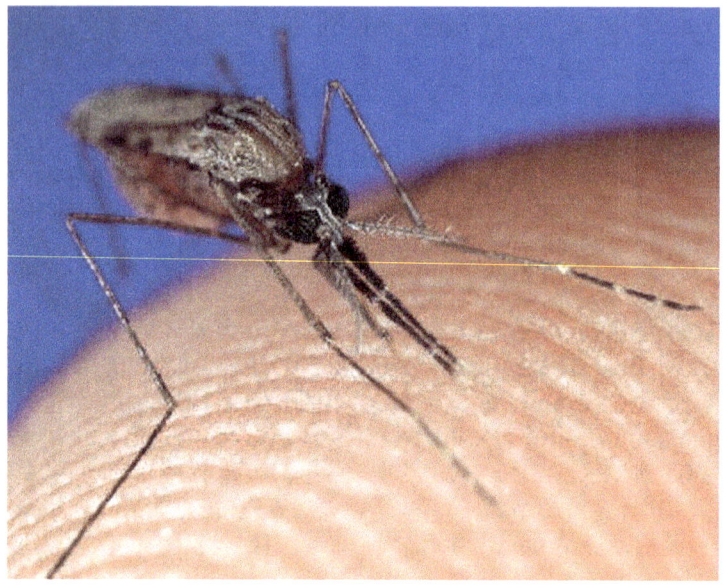

Mosquito biting the skin; penetrating this tough layer requires specialized highly evolved mouthparts.

Although not explicit, size helps distinguish the two. Typically endoparasites are smaller in size than ectoparasites. This makes sense, as in order to parasitize our insides these parasites need to be smaller than we are. They inhabit blood vessels, internal tissues and even individual blood corpuscles, so they are inevitably smaller in size.

Ectoparasites can be larger as they don't need to live within and move around inside of us. So a tick can measure as large as one centimeter across. However, this is only a generalization. As mentioned some roundworms can grow to considerable lengths in the human body; up to three meters long. However, despite this great length they remain extremely thin; they feed through diffusion and each body cell has the greatest surface area possible so that the maximum amount of nutrients can be absorbed from the host.

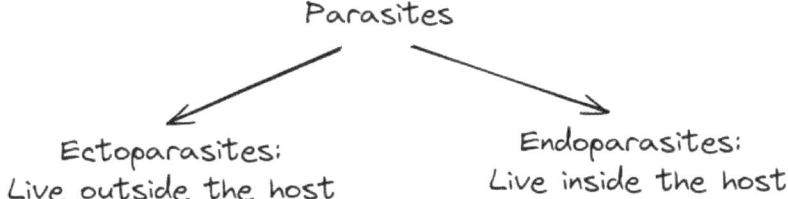

Parasites

Ectoparasites:
Live outside the host

Endoparasites:
Live inside the host

Perfect parasites? How this book is organized
This chapter is entitled 'Perfect Parasites'; this is one that is so highly evolved that it can successfully exploit its host. Ideally the perfect parasite will have developed such a close relationship with its host, that its effects will be minimal, ensuring a renewable resource for generations to come.

This book is organized into sections. The first obvious group of ectoparasites are those which rely upon us as the only host. We are home to these parasites and they require only us. Thus the first section is entitled 'human parasites' and includes those parasites that require us to survive and are somewhat specialized upon us. The next section looks at those which do not live on us, living elsewhere, merely exploiting us when required, mostly for blood meals. These are not specific parasites on us, but often target a range of species of which we are just one. This section is entitled 'blood feeders'. The final section looks at those parasites that parasitize us by accident. I could have tried to organize it differently, maybe using taxonomy, but this is problematic because the parasites cut across groupings not allowing easy classification in this way.

Breaking in: human skin

The skin is the first layer of defense that parasites, or at least ectoparasites, will encounter. So it makes sense to describe it in some
detail. It is also perhaps the principal layer of defense against such
parasites. The skin forms our outermost layer and although it has
many functions, including thermoregulatory and sensory, maybe its
most important function is in protecting us from the outside ele-
ments- and that includes parasites which wish to feast upon us.

We might think of our skin as a fairly useless barrier. After all it
seems pretty soft, flexible and delicate to us. However, if you think
at the scale of an insect it appears much more formidable. Our skin
is only two millimeters thick, that is 0.07 of an inch. However, to
an insect parasite this is a considerable barrier. Even if one consid-
ers a tick, which is a relatively large parasite measuring nearly one
centimeter in length, scaled up to our size our skin is the equivalent
of about two foot deep, 60 centimeters. If you were faced with a
wall 60 centimeters thick would you think it was an easy and thin
layer to get through?

Although we think of our skin as delicate and flexible, this flexibili-
ty actually adds to its strength and durability. The skin is comprised
of a number of different layers each with different properties and a
slightly different make-up. Outermost is a hard and tough keratin
layer. Keratin is the substance which our hair and nails are made of.
Although this keratin layer is thin, it is hard and resilient. Beneath
this is a layer which is known as the stratum corneum; this is basically a thick layer of dead cells, which are however horny and
tough. Any parasite has to penetrate this before it even gets close
to any living material. Below this is the epidermis proper which is
the uppermost layer of skin and the first living layer.
The epidermis has its own obstacles to potential parasites. This layer has numerous nerve endings. This can be compared maybe to a

minefield, or maybe a series of infra-red burglar alarms around a building. A parasite risks 'triggering' any of these while trying to penetrate us. Should it do so, the potential host could potentially realize it is under attack and take action to stop it.

Only with the second layer, known as the dermis, can the parasite find any sustenance. This layer is rich in a network of blood vessels. Should the parasite tap any of these, then it can finally feed. However, having got this far, other problems arise, such as triggering the immune system, and that the blood could start to clot.

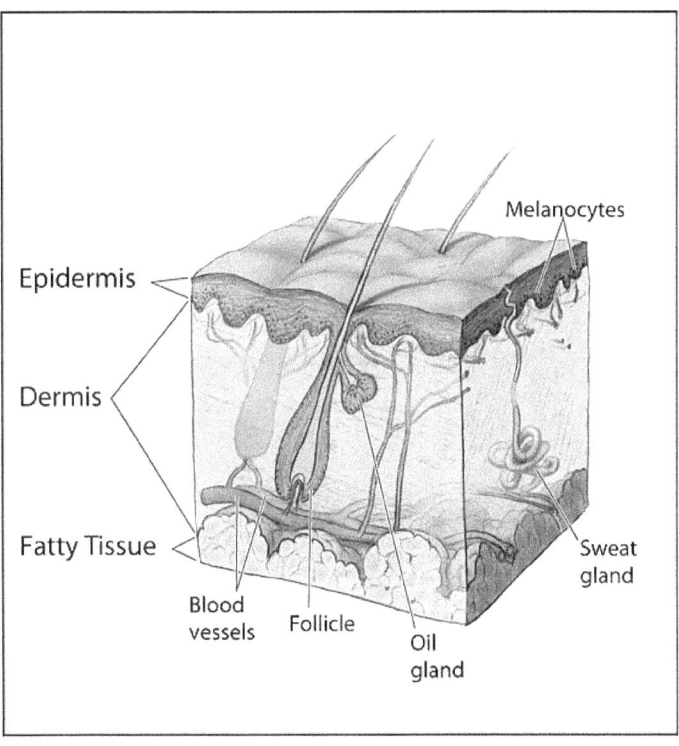

SELECTED REFERENCES

The number of parasites
Carlson CJ, Dallas TA, Alexander LW, Phelan AL, Phillips AJ. What would it take to describe the global diversity of parasites?. *Proceedings of the Royal Society B*. 2020;287(1939):20201841.

What is a parasite?
Although now old, still an excellent starting place:
Price PW. General concepts on the evolutionary biology of parasites. *Evolution*. 1977:405-20.

Academic texts
A good introduction to human parasites:
Cox FE, editor. Modern parasitology: a textbook of parasitology. John Wiley & Sons; 2009.

Introduces parasites from an evolutionary perspective:
Poulin R. Evolutionary ecology of parasites. Princeton university press; 2011.

HUMAN PARASITES

Those parasites that live primarily upon us

2

Bed Bugs

Our beds and bedrooms are a place of safety. We make them warm and comfortable. They are intensely private and personal spaces. We feel comfortable there. That we make them so is understandable; when we are asleep we are probably at our most vulnerable. We need somewhere we can retreat where we know we can be safe.

So maybe nothing is more upsetting than the thought that this intensely personal space will be invaded by revolting and unpleasant parasites. Many parasites are revolting enough, but the fact that Bed Bugs inhabit our most intimate and personal of locations makes them particularly repugnant.

Often those with Bed Bugs will not initially realize their beds are harboring them. The initial signs might be spots of blood on the pillows and bedding; probably just a small wound. Or a fleck of brown on the sheets from their feces, which is thought to be just some dirt. Or a reddish round rash, which is mistaken for a spot. However, before long these parasites are recognized.

Bed Bugs rarely cause any clinical or medical problems. The bites remove blood and can often result in a reddish bite mark on the skin. However, the amount of blood removed is minimal and not a problem itself. But the parasitism can cause allergic reactions in those who are particularly sensitive. At the very least infestation is most unpleasant and the thought of being parasitized most repugnant.

Taxonomy

There are two species of Bed Bug which parasitize humans. Bed Bugs are insects of the Hemiptera order, which are the 'true bugs'. This is a large order, with 80,000 species. These don't undergo a full metamorphosis.

The Bed Bugs themselves belong to the Cimicidae family. The Cimicids are parasitic insects, which feed off the blood of warm blooded animals. These are not only parasites of mammals, Bed Bugs specialized upon Avian hosts can be a major problem in the poultry industry. These inhabit the wood shavings used as bedding or live amongst nest boxes. Altogether there are about 90 species of Cimicids.

The two human species are the Common Bed Bug (*Cimex lectularius*), which is the temperate species and most common species. Whilst in tropical areas the main species is the Tropical Bed Bug (*Cimex hemipterus*).

The Common Bed Bug *(Cimex lectularius).*

Physical appearance

In appearance Bed Bugs are small in size, oval shaped, and do not possess wings. In color they are a reddish-brown. They are clearly visible to the naked eye, measuring up to seven millimeters across. However, although one might imagine they would be obvious, they are often overlooked.

When not feeding they nestle in bedding and under beds, living amongst wooden slats and posts of beds and furniture, only emerging when the host is in bed.

Feeding

It takes about five to ten minutes for the Bed Bug to obtain a full meal. The classic sign is of three bite marks in a row, known colloquially by parasitologists as "breakfast, lunch, and dinner" marks. These are most commonly seen on the exposed parts of the body, such as the face, neck and arms.

For some inexplicable reason, some individuals seem immune to the parasites, or simply not favored as hosts. Thus it is possible for some people in a household to experience signs of parasitism, whereas other lucky individuals equally exposed to them do not. In some individuals who are sensitive to the saliva the bites can become raised due to reaction of the bodies immune system.

The bites of the Bed Bug can be intensely itchy and painful. A common problem for the feeding parasites is that the hosts natural defense mechanism includes coagulant chemicals in the blood which act to thicken blood where any wound occurs, thus preventing excess loss of blood and plugging any gap through which bacterial infection might occur.

Blood sucking parasites overcome this mechanism often by injecting saliva into the hosts; this contains chemicals which act as anti-coagulants, meaning that thickening of the blood does not occur while they are feeding. However, it is chemicals within this saliva which often causes problems for the host. This substance is recognized as being 'foreign', and an active defense response is initiated. It is this which leads to painful itching.

Lifecycle
There are five developmental stages, known as 'instars'. Essentially nymphs molt successively into larger versions of themselves until they reach the final 'adult' stage. The whole process takes about four months.

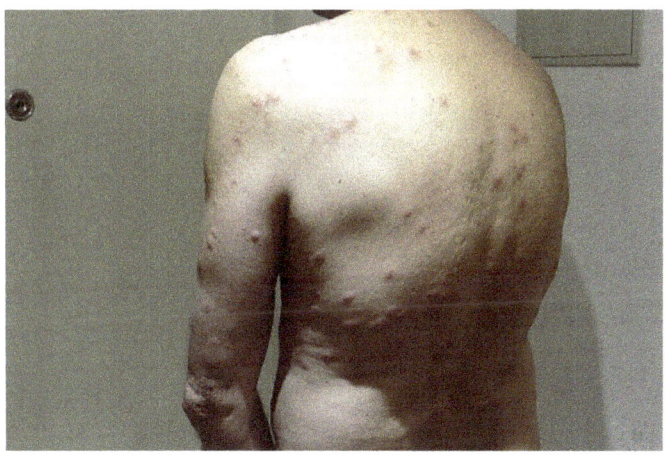

A nasty surprise for a hotel guest!
The bites from Bed Bugs.

Hygiene
Bed Bugs are associated with dirty bedding, mucky living conditions, and old dilapidated houses. However, although such conditions do favor them, they are increasingly being found in more modern dwellings which are well kept. They were once relatively

rare in the developed world, but are becoming increasingly common.

Travel to hotel rooms may be facilitating their spread and dispersal. We are increasingly taking breaks. There is more frequent changing of guests these days, with people taking shorter breaks. Hotel cleaning staff are increasingly being outsourced, or have strict time targets, which means Bed Bugs often survive unnoticed in such locations.

SELECTED REFERENCES

Doggett SL, Dwyer DE, Peñas PF, Russell RC. Bed bugs: clinical relevance and control options. *Clinical Microbiology Reviews*. 2012;25(1):164-92.

Reinhardt K, Siva-Jothy MT. Biology of the bed bugs (Cimicidae). *Annual Review of Entomology*. 2007;52:351-74.

3

Demodectic Mites

Demodectic Mites are usually not considered to be an actual parasite of humans. Instead they are usually considered to be a normal inhabitant of our skin, typically causing no detrimental effects. Biologists classify organisms which live upon, or in close association, with others depending on whether they have an adverse effect on them or not. As such, Demodectic Mites are biologically what is known as a 'commensal'; an organism that is an unobtrusive passenger, dwelling upon us, but not damaging us.

Typically Demodectic Mites inhabit the hair follicles of the eyelashes, where they feed on dead skin material. However, they can be found across the face and forehead. They are common in the animal kingdom; a variety of mammals have types of Demodectic Mites specialized to them.

Taxonomy
Demodectic Mites belong to the Demodicidae family. There are several dozen species, which parasitize a range of mammals. Each species is host mammal specific. For example, the parallel species *Demodex canis*, is seen on dogs. Two species occur on humans. *Demodex folliculorum* and *Demodex brevis*.

A rather neat habitat differentiation appears to have occurred, with *Demodex brevis* inhabiting hair glands and *Demodex folliculorum* the hair follicles themselves. *Demodex brevis* is shorter in size; brevis is Latin for short. *Demodex folliculorum* measures up to 0.4 millimeters across. *Demodex brevis* is considerably smaller at 0.2 millimeters.

Physical appearance

In appearance Demodectic Mites appear to have an elongated cigar shape. The small size means they are barely visible, pretty much invisible to the naked eye. The distinct features they possess only become apparent with the use of magnifying lens.

They are colorless and have no hairs. Both sexes are of a similar size. As is typical for mites they have four pairs of legs which they can use to move over skin rapidly, despite their small size. They are recorded at moving at up to 16 millimeters per hour.

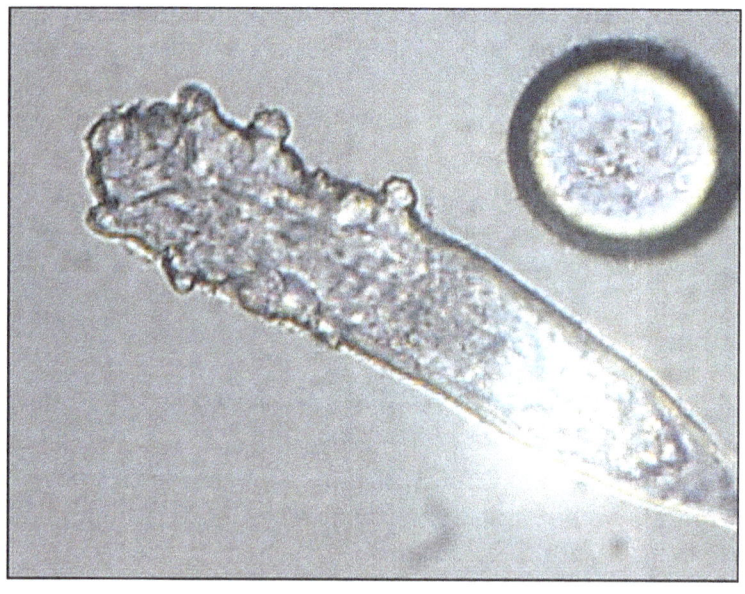

Outline of a Demodectic Mite.

Unlike the Sarcoptic Mite, where the segments of the body are somewhat clearly demarcated, those of Demodectic Mites are ill defined. The frontal body section, known as the opisthosoma, can be more clearly made out as it is striated. The four pairs of stumpy legs all come from this frontal section. The legs end in sets of claws. There are no setae, in other words hairs, on the legs or body.

Lifecycle

As already mentioned Demodectic Mites are found on the face, the forehead, on the cheeks and across the nose. They are also particularly common on the eye lashes. Mites typically inhabit the hair follicles where they live in the deep recesses far away from the surface of the skin. They do not like light, so recede during the day only emerging at night.

The entire lifecycle can be completed within the roots of the hair follicle within 20 to 30 days. Gravid females lay 20 to 24 eggs at the base of hair follicles. These hatch out into a proto-nymphs after about 60 hours. These have a set of three pronged claws on each leg. After another 72 hours these molt into full nymphs which have two sets of claws on each leg. Another 60 hours is required for nymphs to become adults. After maturity is reached adults move out of the hair follicles.

Signs and symptoms

Although they are not actually parasitic, instead sustaining themselves on dead skin, these parasites can still cause problems. This mainly occurs when there is a large increase in numbers. In most people they don't usually cause a problem. However, they can flare up in those who are ill, or the elderly, when the immune system becomes weakened. When this happens Demodectic Mites can increase without check, quickly causing skin problems which can become visibly noticeable.

Many of the symptoms are minor. Mites typically cause a reddening of the skin, with red patches being hard and rough. In mild cases the problem is localized, with several small localized lesions occurring. These are often confined to the nose or around the eyes. They can also cause the rosacea typical of reddened cheeks.

The condition of being infested is known technically as 'demodicosis'. Demodex Mites can be controlled by regular cleansing of the skin and using ex-foliating materials to reduce numbers. Simply using a baby shampoo on eyelashes and associated areas is often enough. However, this in no way implies they only occur on dirty skin. Use of cleansers and moisturizers also helps. Specific treatments such as benzyl benzoate, sulphuric creams or permethrin can also be used in particularly severe cases.

SELECTED REFERENCES

Hom MM, Mastrota KM, Schachter SE. Demodex. *Optometry and Vision Science*. 2013;90(7):e198-205.

Rufli T, Mumcuoglu Y. The hair follicle mites *Demodex folliculorum* and *Demodex brevis*: biology and medical importance. *Dermatologica*. 1981;162(1):1-1.

Delusional Parasitosis

In 1951 an article was published in the Proceedings of the Entomological Society of Washington by esteemed scientist Jay Traver. The article described the authors personal experiences of being infested by mites within her scalp, describing the principal symptoms. No one is doubting Travers integrity; she truly believed she was infested and was experiencing these symptoms. It is now generally acknowledged that she was not infested but was experiencing a mental condition known as 'delusional parasitosis'.

A fear of parasites could well be innate and natural. Perhaps natural selection favored those individuals who found parasites abhorrent, and who went to lengths to avoid them. However, such fears can get out of hand. Delusional parasitosis is the name for the mental condition where sufferers believe they are being parasitized when they are not.

Those affected by this believe that parasites are crawling over them or moving around inside of them, and believe they can feel their movement, despite the fact that there is nothing there. There is a special term to describe this sensation; 'formication'. I guess we have all had that sensation of something crawling over us, maybe the sensation of a wisp of hair or the movement of the air, especially when someone suggests this. But in most cases this is a figment of our imagination and quickly passes. For those with delusional parasitosis it does not pass, instead becoming a long term problem and sometimes obsession.

Delusional parasitosis typically develops when an affected person suffers some trauma associated with parasites. This could be through coming into contact with someone infested, or being bitten or stung themselves. Those affected typically repeatedly seek medical help for the 'parasites'. Common is for them to visit doctors with evidence for the infestation, such as photos or specimens of purported parasites. In fact this has even been named the 'matchbox symptom' as sufferers frequently turn up with a match-box containing a supposed parasite.

Although delusional parasitosis might appear bizarre, it is much more common than one might think. In a review article Suh (2020) found an incidence of two cases per 100,000. The underlying cause of delusional parasitosis are probably chemical imbalances in the brain.

Although delusional parasitosis might appear frivolous and somewhat titillating, the impact of this mental illness should not be underestimated. There has even been an instance of a patient with delusional parasitosis attempting to murder their medical practitioner, the delusion became so overpowering.

REFERENCE

Bourgeois ML, Duhamel P, Verdoux H. Delusional parasitosis: Folie a deux and attempted murder of a Family doctor. *British Journal of Psychiatry*. 1992;161:709-711.

Suh KN, Keystone JS. Delusional Infestation (Delusional Parasitosis). In *Hunter's Tropical Medicine and Emerging Infectious Diseases*. 2020:1132-1136. Elsevier.

Dust Mites

A team of Dutch scientists were responsible for identifying dust mites and naming the genus. The story began in the 1930's when Leiden scientist, Willem Storm van Leeuwen, speculated that mites present in house dust could be responsible for various respiratory allergies. This idea was taken up in the 1940's by Reindert Voorhorst, who was working at the Academic Hospital Leiden.

However, no progress was made. Research was hampered by the war. In the years after the war, Voorhorst and colleagues continued to investigate other mites and the relationship they had with allergies. There was believed to be a connection between the mites found in household dust and storage mites, with it being believed they might be the same species.

In the 1960's Voorhorst enlisted the help of zoologist Prof. Don Kuenen who tasked one of his students, Marise Boezeman, with investigating the mites found in household dust. It was Marise who finally realized that Dust Mites were a separate type of mite, which led to their naming as a new genus; *Dermatophagoides*.

Taxonomy
Dust Mites are a type of Arachnid. Although they rely on us to survive, they are not parasitic. They instead feed on the dead skin material we shed everyday. They have a lifespan of 65 to 100 days.

Physical appearance

When you see an image of Dust Mites they appear quite nasty little revolting things. They are certainly not something you want to think about as being in your house or in your bed.

But we can't see Dust Mites, which maybe makes them appear less menacing. They are minute in size, measuring only about 0.1 to 0.2 millimeters in length. You need a magnifying device in order to see them at all.

Under a magnifying lens they usually appear translucent and see-through. They are obviously mite like in appearance, having four pairs of legs, a round globular body, and the occasional hair which helps them remain attached to our bedding or clothes.

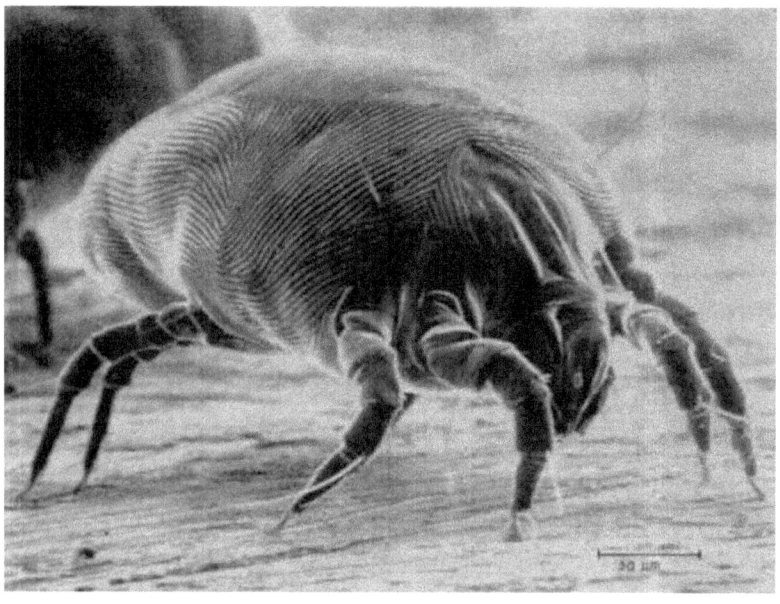

**North American
House Dust Mite.**

Habitats

Our beds are ideal habitats! Firstly, we spend a large proportion of each day in them. This means they contain plentiful amounts of the dead skin matter mites require to live on. Unlike clothing, which is often changed either daily or every few days, bed linen is often only changed weekly or so. This means a lovely build up of dead skin matter occurs; perfect habitat for Dust Mites.

Dust Mites absorb moisture through their exoskeleton. Beds are often moist; we sleep in them, and although imperceptible to us, we are constantly sweating. This makes them moist enough for these mites to thrive in.

Dust Mites and allergies

Our bodies immune system has evolved to recognize foreign substances that might make there way into our bodies, and initiate a response to eliminate them. There are various entry points where such foreign bodies could easily find a way inside of us. These are protected in various ways. Our eyes are covered with a protecting membrane which helps keep bacteria and other pathogens away. Our nostrils are covered with mucous membranes which produce a substance rich in pathogen killing material.

Our immune system has a complicated method of identifying invaders. When it does so, an immune cascade is initiated. Should a foreign substance make its way inside, then our body initiates a range of defense mechanisms; for example our nose begins to run and we sneeze.

For most people, this is not a problem. But some people can be extremely sensitive. Where such a problem is thought to stem from Dust Mites it is possible to have allergy tests. Bad cases can be treated with normal everyday allergy remedies such as antihistamines.

House Dust Mites
(*Dermatophagoides pteronyssinus*).

Minimizing impact

There is not a deal we can do about Dust Mites. They are there, however clean you are. They are just a part of the natural fauna associated with us. For the most part, they don't cause any issue and are not the cause of any disease. The exception is in people particularly sensitive to allergens.

There are, however, ways to minimize their numbers. Bedding should be washed at high temperatures regularly. This eliminates Dust Mites as they can't survive at temperatures above 130 Fahrenheit. For those with allergies, plastic sheeting around the mattress can reduce the amount of Dust Mites, meaning there is less space for them to live. Carpets harbor dust. Wooden flooring or tiles substantially reduce dust, meaning there are less allergens around for those sensitive to them. Special products are available which reduce Dust Mite incidence, for example, De-Mite.

Ways to reduce Dust Mite numbers

The following methods can be used to help reduce Dust Mite numbers and keep them in check:

- Have as few carpets and upholstery as possible. Go for wooden flooring.

- Vacuum and dust regularly.

- Reduce humidity in the home. Use a dehumidifier.

- Wash bedding regularly; Washing at over 50°C kills the mites.

- Remove padded headboards and plush fittings from beds.

- Consider plastic sheeting underneath bedding.

Dust Mites thrive where humidity is high, so dehumidifiers are thought to be of help in their control.

Remember it is no good trying to get rid of Dust Mites. They are simply there and part of the natural ecosystem of our homes.

Palaces have them, as well as the smallest cottage. There presence is no reflection on cleanliness.

SELECTED REFERENCES

General
Hart BJ. Life cycle and reproduction of house-dust mites: environmental factors influencing mite populations. *Allergy*. 1998;53:13-7.

History of discovery
Spieksma FT, Dieges PH. The history of the finding of the house dust mite. *Journal of Allergy and Clinical Immunology*. 2004;113(3):573-6.

Preventing asthma
Gøtzsche PC, Johansen HK. House dust mite control measures for asthma: systematic review. *Allergy*. 2008;63(6):646-59.

5

The Body Louse

In May 1997 the World Health Organization declared an emergency in the small African state of Burundi. Since the beginning of that year more than 24,000 people had contracted epidemic typhus. This became the largest global outbreak in 50 years, and a reminder that such diseases can quickly return if suitable conditions are provided for them.

We might think of typhus as one of those conditions that has been consigned to the history books. Maybe it evokes memories of Victorian slums or wartime concentration camps. Although large epidemics of typhus like that seen in Burundi are now rare, the disease remains endemic in many areas of the world. It is out there waiting to make a comeback. There are a small but notable number of localized outbreaks that still cause a significant problem each year.

There are a number of different strains of typhus. One, 'bush typhus', probably derives its name from the fact that people often contract it after being active in the 'bush' or the outdoors. The vector are mites such as 'Chiggers', of the Trombiculidae family. But rodents can help spread infected mites around. The condition crops up in campers and outdoors workers. But only small numbers of people contract this at once.

The strain of typhus we associate with large outbreaks, and which sweeps through populations, is known as 'epidemic typhus'. The

vector of epidemic typhus is the Body Louse and this parasite
thrives when human sanitation is compromised.

Introducing the lice

The Body Louse is one of three types of lice inhabiting humans.
The lice are a large and successful group of insects. There are more
than 5,000 species. Traditionally taxonomically lice were placed in
the order Phthiraptera.

Species within the lice group are characterized by being wingless,
possessing a body comprised of a head, thorax and abdomen, and
having three pairs of legs. Traditionally there are two suborders of
lice; and species were placed in them depending on the anatomy of
their feeding apparatus.

The Mallophaga are the 'chewing lice', with mouthparts designed for
breaking up and chewing on food. There are about 3,000 species of
chewing lice. These are parasites mainly of birds, and also some
mammals. Those keeping poultry are sadly well aware of the
problems caused by such parasites. For example, the Red Mite
(*Dermanyssus gallinae*) is a common parasite on chickens and turkeys,
and can remove significant quantities of blood from affected birds
leading to a considerable economic impact.

The other grouping of lice are the Anoplura, known commonly as
the 'sucking lice'. There are about 500 species within this grouping.
These have highly specialized mouthparts designed for sucking.
There is a suction tube for blood extraction known as a stylet,
structures to pierce the skin, a suction pump to remove blood, and
another to insert saliva (containing anticoagulants) into the host.

Those lice that parasitize man, the Body Louse, the Head Louse,
and the Pubic Louse are all Anoplura 'sucking' lice. The Head Louse
and Body Louse belong to the same genus, *Pediculus*, while the Pubic

Louse is somewhat different and placed separately in the genus *Phthirus*. Taxonomists sometimes had slight arguments as to whether the lice are an order or a superorder, and some considered the Mallophaga and Anoplura as orders in their own right, but generally they were considered as suborders of the order Phthiraptera.

However, a spanner was put in the proverbial works by genetic analysis. This has painted a somewhat different and rather messy picture. Recent research suggests that the human lice are more closely related to and possibly derived from Bark Lice and Book Lice, which belong to a totally different order altogether. This is despite the fact that they do not appear similar in appearance to these species at all.

Hence modern taxonomy now often places them amongst the Psocodea order, which are the 'Book Lice'. Book Lice, as you might have guessed, are found amongst old books, principally amongst the bindings.

This new genetic evidence means that now human parasitic lice have been relegated to being a mere family, known as the Anoplura, amongst the suborder Troctomorpha, of the Psocodea order. This being decided on the number of segments on the antenna.

Whether this will remain is uncertain, as taxonomic revision occurs periodically. Taxonomists make a living from classifying and re-classifying. Taxonomy is somewhat of a pedantic science, it is best not to worry unduly as to where the human lice as a whole belong as revision is sure to happen again as more genetic analysis is performed and a clearer picture of the true relationship amongst these insects is gained.

The three human lice

There has been much speculation scientifically on the evolutionary history of those lice that parasitize humans. The Body Louse and Head Louse are very similar in appearance to each other and obviously closely related.

The Body Louse is slightly smaller than the Head Louse. They inhabit different parts of the body; as the common names suggest the Head Louse is found predominately on the scalp, where it nestles amongst the follicles of the hair, but it is also found on the eyebrows and neck.

In comparison the Body Louse is found on the rest of the body, or at least will feed from it, as it actually spends most of its time off the human host and amongst clothing. As mentioned these two lice are obviously closely related. Whether they should be considered as separate species in their own right, or whether they are simply variations on the same species is an argument that only taxonomists can get passionate about.

They are thought to be capable of interbreeding, if forced artificially, but on the host will not do so. Today they are generally considered as being variations or different varieties of the same species known as *Pediculus humanus*. The Body Louse being recognized as a variety known as *Pediculus humanus humanus* or *Pediculus humanus corpora* and the Head Louse as *Pediculus humanus capitis*.

Genetic analysis provides an indication of which came first, and suggests that the Body Louse evolved from the Head Louse. This makes sense. Head Lice are happy simply amongst hair follicles, but the Body Louse prefers to live mainly amongst clothing. The Body Louse could only evolve once we started to wear clothes.

David Reed, curator on mammals at the Florida museum of natural history, has used this knowledge to infer when humans first began to wear clothing. Our species beginning to wear clothing opened up a new potential niche for lice parasites, and provided the opportunity for the evolution of more than one species of parasitizing louse species.

Genetic analysis suggests that Body Lice and Head Lice split between 70,000 and 180,000 years ago, providing a likely date for the evolution of clothing in humans. This is notably before humans are thought to have left Africa and traversed the world; when we left east Africa, we were already clothed.

The Pubic Louse, as is apparent, is noticeably different from Head and Body Lice, and is thus placed in its own separate genus. It is thought to have evolved from parasites present on gorillas and have subsequently have moved onto humans. Genetic analysis suggests the human Pubic Louse split from its gorilla cousin about three million years ago. Human are perhaps unfortunate in being parasitized by three separate species of lice, when most host species have only one.

History of discovery

It has long been known that humans had lice; these are a most visible parasite, and the sensation of being infested can hardly be missed. Lice were well known in antiquity. However the first real scientific study of them began with the development of the magnifying lens which allowed their biological features to be examined in detail for the first time.

Famous for his early use of the microscope was Dutchman Jan Swammerdam (1637 to 1680), who drew detailed illustrations of human lice. However, by an extraordinary piece of bad luck, he seemed to have only found female Head Lice to study. He thus falsely concluded that they were hermaphrodites and capable of reproducing without males.

This false conclusion was corrected by later scientist Van Leeuwenhook who managed to find and identify males. In his studies he maintained lice in his stockings, which allowed them to take regular blood meals yet kept them together in an easily accessible place. Through observation of populations, he was able to pretty much determine the course of the entire lifecycle.

Physical appearance
Body Lice are superbly adapted to parasitize us. In form they are well suited for crawling amongst hair. They are somewhat flattened in profile, with tapering elongated and flattened bodies, ideal for crawling between hairs and close to the bodies skin.

These are blood sucking insects, and they possess a sharp stylet to extract regular blood meals, which they do several times each day. Salivary glands produce saliva which is pumped into the site of the bite to prevent the blood from coagulating. When not in use the stylet is retracted into the head and out of the way.

As is typical for lice, the Body Louse has a clearly segmented body, with a head, thorax and long tapering abdomen. They are easily overlooked being only the size of a sesame seed. They often nestle in the lining of clothes between the stitching, where they are both difficult to locate and difficult to remove. Thus people can harbor an infestation for a considerable length of time before they are identified.

The six legs have strong claws which aid attachment to the human host. Males are somewhat smaller than females in size; the females lay eggs, thus are larger to accommodate this.

Photograph of a female *Pediculosis humanis* louse.

Where Body Lice live

A human host is required for survival. Actually, for the majority of the time Body Lice are not found actually upon the person harboring them. Instead they spend most of the time in their clothing, moving only periodically to the skin in order to feed. The females lay eggs in clothing. These 'nits' appear a bit like dandruff and are easily overlooked. Body Lice are most commonly found around the neck and shoulders, under the armpits, and around the waist and groin areas. They are typically found in the clothing associated with these locations.

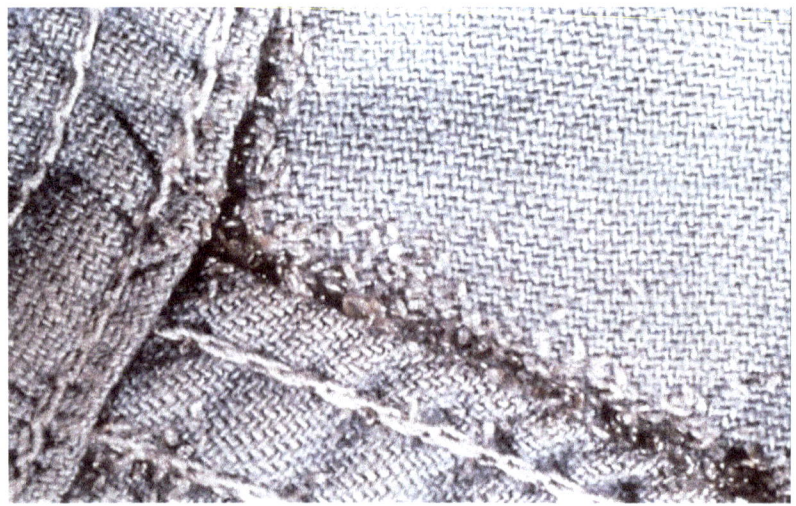

Body Lice live amongst clothing seems.

Lifecycle

Body Lice require a human host for survival and quickly perish away from a suitable host. After only one or two days without a blood meal they will die. Body Lice are fussy regarding their preferred requirements. They like warm temperatures, and have a preference for temperatures in the range of 29 to 32 centigrade. Once temperatures pass 50 centigrade they are unable to survive and quickly die. Adult Body Lice live for approximately 35 days. The females are capable of laying 200 eggs. Thus, a single louse can rapidly multiply into many thousands.

Today, infestation with Body Lice is somewhat rare in the developed west. Where they do occur, it is amongst the homeless, those fleeing war, or those living in temporary accommodation following a natural disaster. Because they live amongst clothing, they occur and thrive where it is not possible to change and wash clothing on a regular basis.

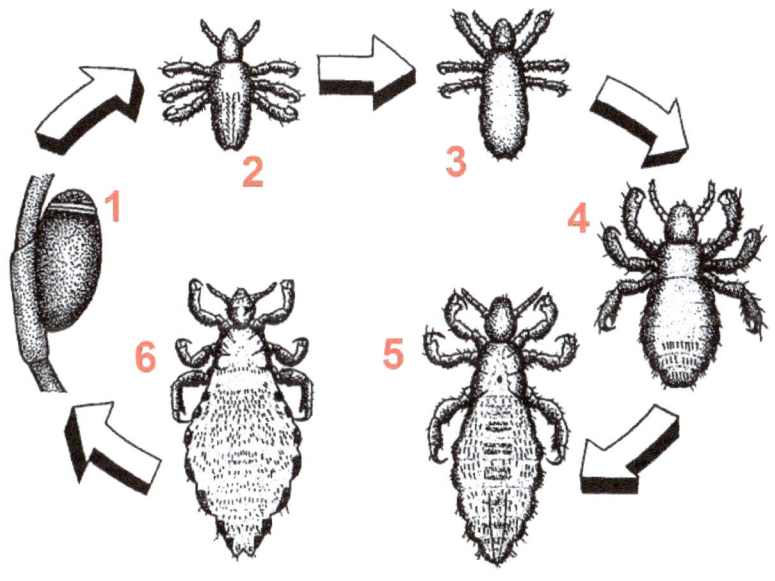

Body Louse lifecycle:
'nits' (1) hatch into smaller instars (2-4),
which eventually molt to become adults (5,6).

The problems Body Lice cause

Actually the direct consequences of Body Lice infestation are relatively minor. They can cause irritation and itching, with those heavily infested scratching profusely. The wounds they cause while feeding can mean spots of blood are present. This is one of the ways to diagnose them, as due to their small size they are often overlooked.

Rashes are also common in those infested. These typically occur around the site where Body Lice are found, but can develop more generally. However, although the direct consequences of parasitism are minor, the problem is that the Body Louse is a vector for a variety of conditions which can seriously impact host health.

Typhus

Typhus is the worst condition transmitted by Body Lice. Those experiencing this illness typically experience symptoms of an acute fever, with headaches and chills. In the pre-antibiotic era the rate of death was high. Today, with antibiotic treatment the rate of survival is much higher.

Disentangling the cause for the disease was difficult. Throughout history mankind has suffered from fever causing illnesses. These illnesses could be caused by a variety of pathogens. How can you determine what was causing a particular bout of fever? Thus distinguishing typhus from other types of feverish illness, then determining the causative agent was difficult. It was just one of a number of similar illnesses.

Typhus is a bacterial disease, belonging to the Rickettsia genus group. There are actually a number of bacteria, each causing somewhat distinct illnesses. However, all tend to cause feverish illness of a recurrent nature, whatever the exact pathogen causing it. Thus in research on typhus there were a number of challenges.

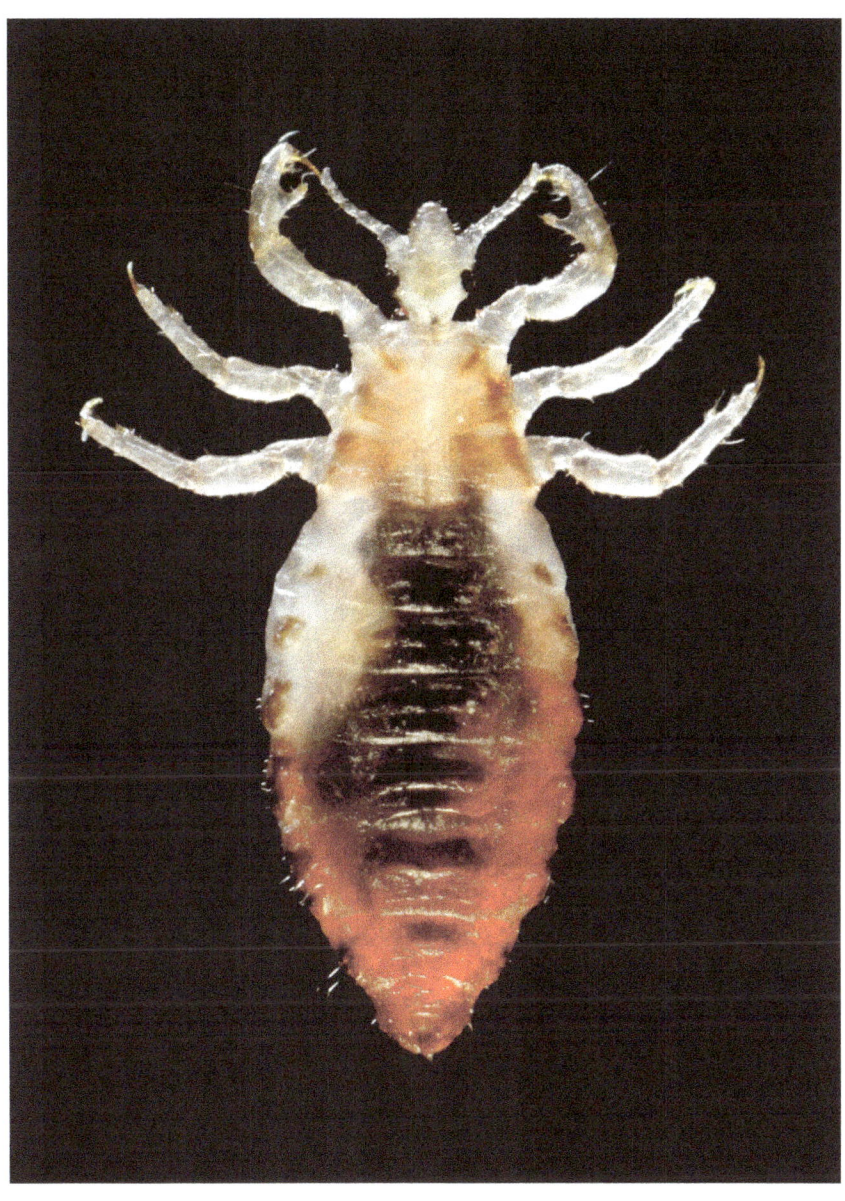

Male body louse. The distal tip of the male abdomen is rounded, whereas in the female it is concave.

First ascertaining what the agent causing the disease was, and secondly determining what was transmitting it.

It did not pay to be one of the early pioneers of typhus research. US Microbiologist Howard Taylor Ricketts studied spotted fever in the early 1900's and identified the causative agent of Red Fever as *Rickettsia rickettsia*, but he caught typhus in the course of his studies and subsequently died of it. One of his researching successors, Stanislaus von Prowazek, also died while trying to replicate his observations in 1914.

The first to name the bacteria responsible for typhus itself was the Brazilian researcher Henrique da Rocha Lima, who to honor his unlucky forebears named it *Rickettsia prowazekii*.

But typhus is a vector borne condition. The next step was determining the vector of the condition. Charles Nicolle discovered that the epidemic form of typhus was transmitted by Body Lice in 1909. His discovery lead him to be awarded the Nobel Prize in 1928.

Epidemics
Naturally typhus is endemic amongst human populations, occurring at low levels. Small numbers of people would be affected, but the condition could not spread because there would not be enough Body Lice around.

For a typhus epidemic to occur there needs to be an increase in the number of Body Lice within a human population, accompanied at the same time by the occurrence of the typhus bacterium in people. Reservoirs of human infection occur. These act as the seeds for future epidemics. It just needs Body Lice to be present in sufficient numbers and density for the typhus bacteria to begin to spread.

Such epidemics of typhus occurred on a regular basis throughout history. Notably in Russia, where waves of typhus would occur at regular periods. When we think of typhus we also often think of Victorian England too. Although actually major outbreaks occurred in 1817 to 1819, 1826, 1827, and 1832, all just prior to the Victorian age. As the opening paragraph illustrated, outbreaks can still occur in the modern day, with the one in Burundi in 1997 being the largest in recent times.

However, maybe it is in times of warfare that typhus and Body Lice can really have an impact. Not only are there large numbers of fighting soldiers, often living in highly unsanitary conditions for extended periods of time. But there is often civilian displacement, refugees, and much movement of people provide ideal conditions for the Body Lice to thrive.

The World Wars stand out as the conflicts we would expect typhus to be a major problem. Perhaps surprisingly, there was no outbreak of typhus on the Western front during World War One. This is despite Body Lice thriving in the trenches. Soldiers in unsanitary conditions, wearing the same clothes weeks on end provided an ideal situation for the lice. However, although typhus did not occur, another condition known as trench fever, which is caused by a bacterial infection (*Bartonella quintana*), did. On the eastern front however, typhus did occur and is thought to have killed 2.5 million in total.

During World War Two there was widespread fear of typhus. This was used by the Nazis to justify rounding up Jews and placing them in Ghettos. Jews were widely blamed for harboring typhus and promoting its spread amongst the general population. However, this was obviously unjust and was only used as an excuse to round Jews up and separate them from the general population. By doing this Jews were then being placed in exactly the situations where the

49

disease could thrive.

That is what one would expect anyway. But research has examined typhus in the Warsaw Ghetto and modeled what should have occurred given the large amount of overcrowding Jewish people were forced into. Expected would be a large surge in typhus numbers as the Ghetto became more and more crowded in the 1940's.

But actually in real life numbers fell! This was thanks to the vigilance of the (Jewish) medical authorities in the Ghetto. Completely contrary to Nazi assertions that Jews spread typhus the Jewish population proved well capable of managing the risk of it, despite the difficult situation they had been placed in. This was proof enough that Jews were no harbingers of this disease.

Overleaf: Russian poster warning of Body Lice. Waves of epidemic typhus were common in Russia. This poster dates from 1921 and warns Red Soldiers of the dangers of the Body Louse.

SELECTED REFERENCES

Biology/General
Mathison BA, Pritt BS. Don't Be a Nit Wit; Know Your Lousy Companions!. *Clinical Microbiology Newsletter*. 2022;44(13):115-22.

Epidemic typhus in Burundi
Raoult D, Ndihokubwayo JB, Tissot-Dupont H, Roux V, Faugere B, Abegbinni R, Birtles RJ. Outbreak of epidemic typhus associated with trench fever in Burundi. *The Lancet*. 1998;352(9125):353-8.

Evolution of lice
Reed DL, Allen JM, Toups MA, Boyd BM, Ascunce MS. 10 The study of primate evolution from a lousy perspective. Parasite diversity and diversification: Evolutionary ecology meets phylogenetics. 2015:202.

Evolution of clothing
Allen JM, Worman CO, Light JE, Reed DL. Parasitic lice help to fill in the gaps of early hominid history. *Primates, Pathogens, and Evolution*. 2013:161-86.

Typhus control in the Warsaw Ghetto.
Stone L, He D, Lehnstaedt S, Artzy-Randrup Y. Extraordinary curtailment of massive typhus epidemic in the Warsaw Ghetto. *Science advances*. 2020;6(30):eabc0927.

6

The Head Louse

Nits! When at school the rumor that so-and-so had Head Lice and nits was damning. It was an unjustified and untrue suggestion of uncleanliness and lack of hygiene. Revolting.

Head Lice are perhaps the most obvious of the parasites that infest our children. The part of the body they inhabit perhaps makes them seem worse. Our heads are our most prominent feature. Not only that, the head is the part of the body which is exposed to the rest of the world. The fact that these parasites occur on the part of the body we most associate with our personal persona makes these parasites particularly disturbing. The incidence at which Head Lice occur in the Western world has increased in recent years; up to 30 percent of children being thought to be affected at any one time.

Physical appearance
The Head Louse is similar in appearance to the Body Louse described in the previous chapter. They are obviously closely related. There continues to be debate as to whether they can be considered the same species, or as a distinct variety. They are currently generally recognized as being a distinct variety. The principal physical difference with Body Lice is in size; the Head Louse is somewhat smaller. However, even so considerable overlap occurs. Otherwise they are fairly similar anatomically.

As is characteristic for insects, the Head Louse possesses three pairs of legs. The body follows the insect bauplan comprising a head, thorax and abdomen. The abdomen is elongated and heavily segmented. They are wingless and thus unable to fly, but this does not mean they do not move. They can crawl rapidly and are highly mobile. Flight is an expensive and dangerous luxury. Also the sensory organs such as eyes are reduced in size. They are not needed in order to locate new hosts.

Male (top) and female (below) human Head Lice,
(*Pediculus humanus capitis*).

Lifestyle: Use your head

The medical term for infestation with Head Lice is pediculosis capitis; essentially louse infestation of the head. Where Head and Body lice do differ is in their behavior and way of life. The common names underline this fact. The Head Louse is found mainly on the head, although it can also found on the eyebrows. They can also be found on other parts of the body on occasion. Although they will move off a person, they are not able to survive for any length of time off a human host. This is in contrast to the Body Louse which is happy to dwell amongst our clothing and the fabrics which come into close contact with us.

The difference in lifestyle is most obvious in terms of where the 'nit' eggs are laid. The Head Louse lays these near the base of shafts of hairs on the scalp, whereas Body Lice lay theirs off the human host and in the seems of clothing.

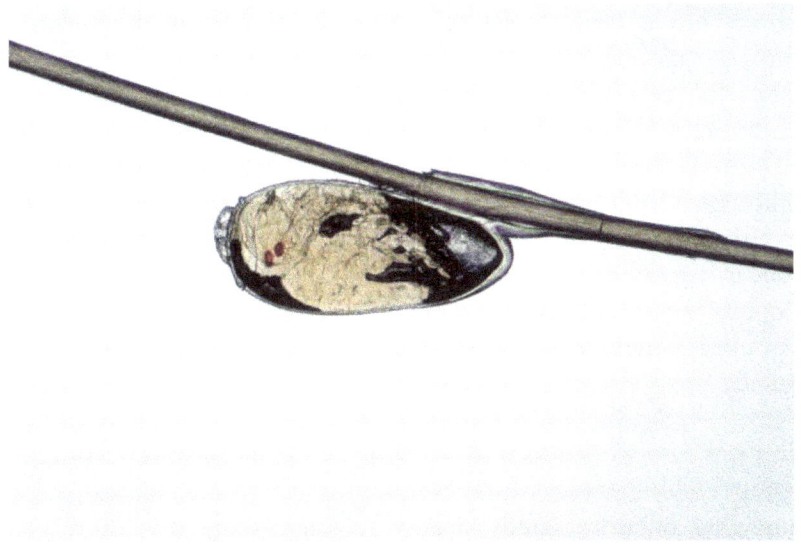

Unhatched nit of the parasitic Head Louse.

The full lifecycle lasts about 35 days. Nits are small in size, being about the size of a pinhead. They are immobile and remain stuck to hair. They are easy to overlook as they are often mistaken for dirt or dandruff. Nits which are 'alive' have a darker spot. Once the nymph has hatched and emerged, the shell of the nit remains, often with an open ending where the nymph has left.

Nymphs emerge from the nits about seven days after being laid. Nymphs are minute, not much larger than the nits from which they emerge. They are essentially miniature versions of the adult, resembling them in appearance. Nymphs go through a succession of stages, molting into a series of larger versions until they reach the adult size. They undergo three molts and reach the adult size after about only a week.

Adults are only the size of a sunflower seed at most. Thus it is easy to understand why they are often overlooked. The adults live on the human head, feeding several times a day from the scalp by removing blood. Females are larger than males; it is the females which lay eggs. The females can lay up to eight eggs daily if conditions are favorable.

Although mobile in the sense of crawling rapidly, Head Lice nevertheless typically require some contact between people for them to spread. However, they can be transmitted between people in combs and towels which come into contact with head hair. Most commonly they move between people through head-to-head contact.

Thus it is easy to understand why infestation is common in children. Young children lack awareness of physical presence and remain somewhat uncoordinated. Heads frequently bang together in the classroom or in physical play, which offers the opportunity for Head Lice to move around. We can all remember playground tumbles. Typically, such play reduces as we get older.

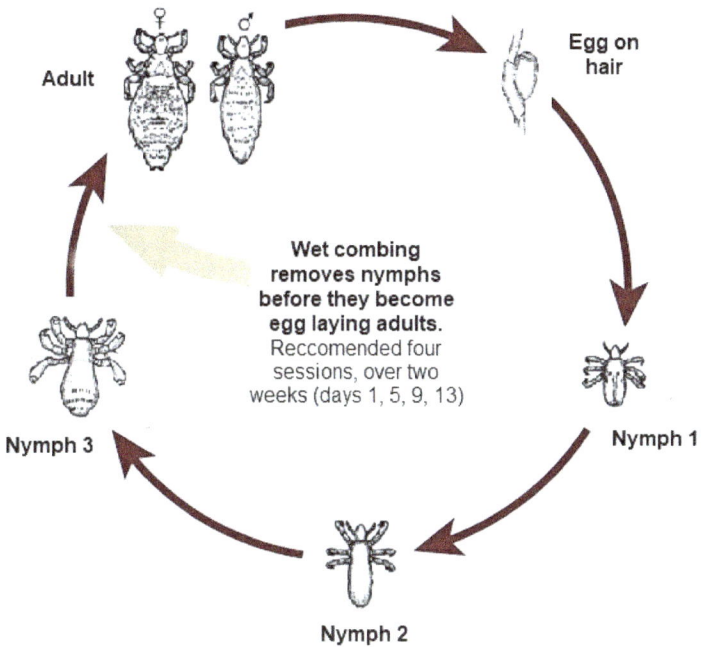

Wet combing removes nymphs before they become egg laying adults. Reccomended four sessions, over two weeks (days 1, 5, 9, 13)

The Head Louse lifecycle.

The age range of children most commonly infested are those between four to 11 years of age. Although it is a perhaps a natural instinct to subconsciously associate infestation with dirtiness or uncleanliness, this is incorrect. It is often stated that the opposite is the case with Head Lice having a preference for clean hair. This idea was probably propagated in order to give comfort to those whose children were affected. Actually, it appears that Head Lice show no preference whatever for clean or dirty hair; they will parasitize whatever head they are lucky enough to land on.

It is perhaps a distressing irony that actually it is girls, who arguably take more pride in their appearance, who are more likely to become infested than boys. This is because girls have longer hair, providing more space, and more cover to hide in, than is the case for boys. One can also well imagine an older girl becoming infested from a younger male sibling.

How common are Head Lice? The prevalence of Head Lice, the proportion of the population affected by them, appears to have increased in recent decades. Traditionally a range of chemical treatments were used against Head Lice, but increasingly resistance to these has developed. Typical parasitic loads can be quite small, with individuals harboring about a dozen lice in total. However, larger loads can occur with particularly unlucky individuals harboring many more.

However, it would be a mistake to think that Head Lice are exclusively a problem for children. The parents of children who harbor infestations, and adults who care for them, will also routinely become infested simply due to the close proximity they have with such children.

Also frequently affected are adult carers, such as nurses or those who care for the elderly, the homeless, those who are infirm or adults with learning issues. Such groups can often have Head Lice due to their lack of self-awareness and frequent head contact that that entails.

It was common to de-louse historically.
Jan Siberechts, Cour de ferme. 1662.

Are you a nit wit?

It is rare for infestation with Head Lice to cause any clinical problems. However, infestation can lead to itching. This is not necessarily caused by the actual presence of the lice themselves but rather due to the biting and removal of blood which can cause itchiness. Sometimes there can be skin inflammation; the body attempts to heal itself and an immune response is initiated which can cause itching. Such reactions can be quite severe, but this is rare.

Although infestation only rarely causes a medical problem, the influence this parasite can have should not be underestimated. The itchiness caused can result in lost sleep and loss of concentration which is detrimental to learning and general well being.

Where infestation is limited and short term, such discomfort does not cause any long term problem. However, where infestation occurs over a prolonged period of time, or where it is continually repeated in nature, the effects can be more significant. In areas of the developing world this can be a major problem and the hindrance caused to educational achievement very real.

How are you feeling today? 'Lousy'. The origins of this commonly used phrase probably originates from infestation caused by Head Lice. It illustrates how the itchy uncomfortable sensation caused by these parasites, and consequent irritation caused, can make us feel. We all know how miserable even minor, but persistent, itchiness can make us.

Another phrase originating from infestation with these parasites is to be a 'nit wit'. We all have an image of a rather dim witted slow child.

Controlling Head Lice

Various traditional remedies have been used to control Head Lice. These include using treacle or olive oil. The principle behind these remedies is that they reduce the mobility of Head Lice, trapping them, and preventing them from continuing their lifecycle. Whether having these substances on your head is socially desirable though, is another matter.

A variety of insecticidal treatments were developed to control Head Lice. These kill the lice, so would appear a good remedy. But you can't outwit, a Head Louse nit. Resistance to treatments rapidly developed. This is an understandable consequence of natural selection; a small minority of lice survive each round of treatment and in turn go on to breed, spreading resistance amongst the remaining Head Lice. Over time treatment becomes gradually less effective. Evidence of resistance to the commonly used Permethrin treatment of the 1990's soon began to be reported. Some resistance has been reported even to the more modern Melathion.

Attempts to slow the development of such resistance were made by advocating treatment of only a single drug, with this alternating with another every few years. This slowed the development of resistance. However, in the late 1990's, in the spirit of promoting free competition, the market was deregulated and consumers were free to purchase whatever product they chose. A biologist could not have conceived a better way to selectively favor resistance. More modern treatments instead work by disabling Head Lice and hindering their mobility. These have proved more effective, as they do not cause the same level of selection and thus resistance to develop.

Combing to remove lice is a traditional method to combat Head Lice and has became more and more popular, especially as the chemical treatments have become less effective. Parents can actually feel they are doing something, which is not the case with chemical treatments. The development of specialized combs aiding removal soon began to be developed commercially. These combs have tiny teeth, at a distance apart ideal for teasing mobile Head Lice out. These often now known as 'bug busters.' Generally there is agreement that use of these combs helps eliminate Head Lice, however technique and persistence are key.

A number of commercial Head Lice combs exist.

The combing procedure is lengthy, and is required on multiple occasions. It's a bit tedious. Best appears to be to use combing in conjunction with some other treatment. But parents seem to prefer combing rather than using chemical treatments.

Long before the advent of modern drugs parents were using a range of natural remedies, mainly of what they had at hand. Currently favored are olive oil, essential oils and tea tree oil. Although these may hinder lice, they are not effective treatments. Added to which, going to bed with a head covered in olive oil is hardly desirable.

How common are Head Lice?
Surveys suggest that in the 1980's and 1990's between two and ten percent of children had Head Lice in the UK at any one time. However, in the 1990's they appeared to be an increasing problem. This was probably due to rising levels of insecticidal resistance. COVID-19 seems to have been a massive check on Head Louse incidence. Practically overnight schools were closed and social interaction prohibited. The transmission channels Head Lice relied on thus disappeared. Evidence from internet search records reflect these changing fortunes. There was a sudden decline in Internet interest in the Spring of 2020.

Much higher levels of prevalence are routinely observed in the developing world. A survey by Falagas et al. (2008) found that prevalence levels of 60 percent were not unusual. What is thus a relatively minor problem in the developed West, is a major cause of educational under achievement in developing areas of the world.

Head Lice can be used to spot cases of neglect and abuse. Where infestations are heavy and go untreated this can be an important indicator of a problem. Those in safeguarding roles can use the presence of Head Lice as an indicator of a potential problem. In some cases infestation with Head Lice has been used as evidence of such neglect.

Head Lice typically lay the 'nit' eggs close to the shaft of the hair close to the scalp. The hair grows from the base, with the hair growing longer with age and the oldest hair being pushed further from the scalp with time. The position of nits on the hair can be used to estimate how long the infestation has been going on for, and how long it has gone untreated, thus indicating the length of child neglect.

In a novel study, Lambiase and Perotti (2018) used this idea to estimate how long an elderly person had been neglected for prior to their death. The person was found to be harboring a severe infestation of Head Lice. The person had been taking treatment for high blood pressure, and it appears that due to neglect had been overdosing on the drug used to treat it. This had had a knock on consequence on the Head Lice, which had been inhibited from laying eggs. The researchers were able to ascertain when the neglect had begun by examining the length of hair where nits had not been laid, and estimated that neglect had been carrying on two months prior to death.

REFERENCES

General review
Walker MD. Head lice suck! Characteristics of infestation and recommended treatments. *Dermatological Nursing.* 2023;22(1).

Epidemiology in UK
Downs AM, Harvey I, Kennedy CT. The epidemiology of head lice and scabies in the UK. *Epidemiology and Infection.* 1999;122(3):471-7.

Downs AMR, Stafford KA, Coles GC. Head Lice prevalence in school children and insecticide resistance. *Parasitology Today* 1999. 15

Walker MD, Sulyok M. Internet searching on the head louse in the UK since the COVID-19 pandemic. *Pediatric Dermatology.* 2023;40(1):96-9.

Head Lice Worldwide
Falagas ME, Matthaiou DK, Rafailidis PI, Panos G, Pappas G. Worldwide prevalence of head lice. *Emerging Infectious Diseases.* 2008;14(9):1493-4.

Use to spot child abuse
Lambiase S, Perotti MA. Using human head lice to unravel neglect and cause of death. *Parasitology.* 2019;146(5):678-84.

Treatments
Mumcuoglu KY, Pollack RJ, Reed DL, et al. International recommendations for an effective control of head louse infestations. *International Journal of Dermatology* 2021. 60(3): 272-280

7

The Pubic Louse

The third type of louse to inhabit humans is the Pubic Louse (*Pthirus pubis*). In some respects this can be considered as being the forgotten of the three human louse species. It is little studied compared to the other two human lice species, and as a consequence less is known about it. It is also less discussed. This is probably because of the nature in which these parasites are transmitted, which without doubt is sexual or at the very least involves contact of a highly intimate nature between people. Pubic Lice are certainly not to be discussed in polite society or at the dinner table!

However, another reason that the Pubic Louse has been neglected compared to other human parasites is that unlike the Body Louse it is not a disease bearing vector. Also to bear in mind is that the location of its parasitism; predominately the groin and lower abdominal area does not facilitate study. Those affected may be reluctant to seek medical attention due to the sexual nature of the mode of transmission and thus the stigma of infestation.

Infestation with Pubic Lice is commonly referred to as having 'crabs'. The Italian renaissance scientist Franceso Redi is probably to thank for this epithet. This scientist is best known for discovering that, contrary to prevailing opinion, flies could not self generate from rotting meat or feces. Instead he showed that they came from maggots, which in turn originally came from eggs laid by flies. Redi seemed interested in a wide range of revolting insects and parasites, for example corresponding with another renaissance scientist, Giovanni Cosimo Bonomo about Sarcoptic Mites and scabies. So an interest in the Pubic Louse was pretty much par for the course for him.

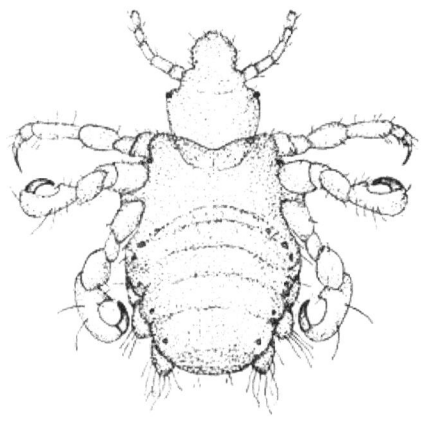

The Crab-Louse (*Pthirus pubis*). Adult male.

**The outline of a Pubic Louse is most distinctive.
Here an adult male.**

Redi noted the crab like appearance of these parasites after examining them under the microscope and as a consequence named them 'crabs'. If examined close up, the parasites do indeed look rather crab like in appearance, with the legs sprayed out to the side.

Taxonomy
The Pubic Louse is distinct enough from the other two species of human louse to be placed within its own genus; *Pthirus*. That Pubic Lice have been placed in their own genus possibly reflects that they had separate evolutionary origins from the other two lice species. The *Pediculus* lice appear to have evolved on chimpanzee ancestors,

while *Pthirus* appears to have evolved on gorillas. The two types evolved approximately six to seven million years ago corresponding to the evolutionary divergence of the two host species. *Pthirus* lice appear to have leapt onto humans three to four million years ago. The Gorilla is parasitized exclusively by the louse species *Pthirus gorilla*. However, as described previously genetic analysis is suggesting a different picture.

Physical appearance

The common popular name of this parasite, the 'crab' louse, is an apt moniker. It does superficially look like a miniature crab in appearance. It has a short broad squat appearance. Sharp claws are present on the end tarsal segments of the three leg pairs. These claws have a serrated surface allowing easy attachment to smooth surfaces and rapid movement.

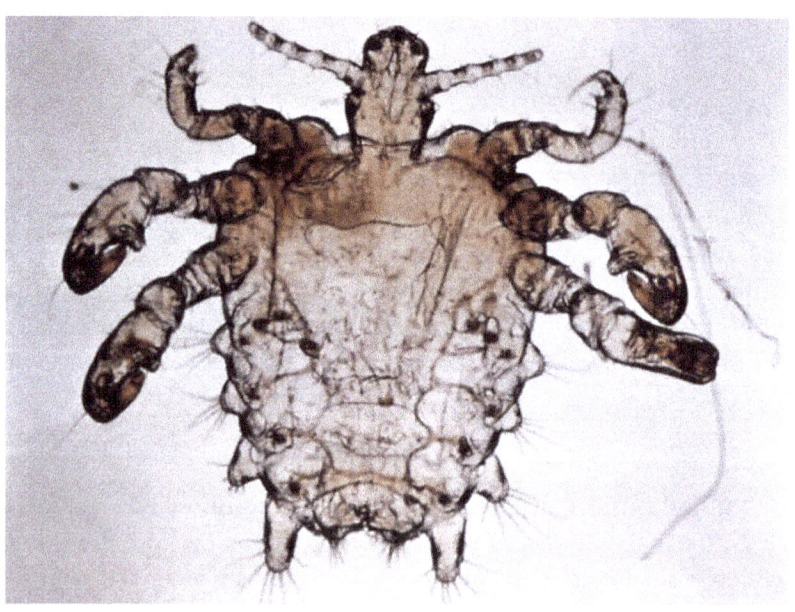

The distinctive outline of a Pubic Louse.

Unlike the other human lice which have leg pairs which are of approximately equal size, those of the Pubic Louse vary in size. The frontal pair are smaller than the rear two pairs, which are much thicker and sturdier. These unequally sized legs enhance the crab like appearance.

Pubic Lice are smaller in size than the Head and Body Lice, measuring approximately one to two millimeters in length. The antennae are shortened and the compound eyes are reduced in size; this parasite does not need sensory perception as it rarely leaves the host. Normally in color they are a brownish gray, but when engorged with blood they take on a red coloration.

The claws are excellently designed for holding tight to host hairs, with sharp pointed tips.

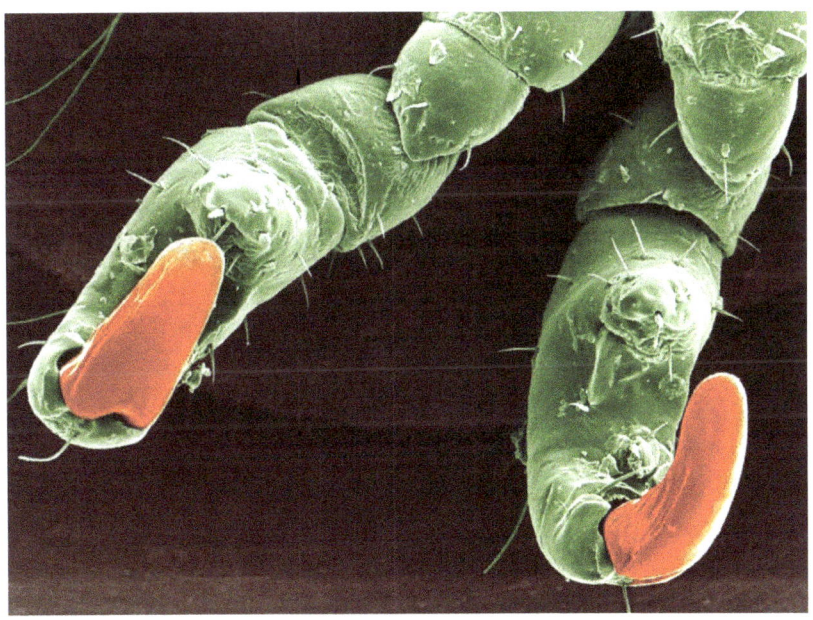

Close-up of the claws of a Pubic Louse.

Lifestyle

As the name logically suggests, the Pubic Louse lives predominately amongst the hairs of the genital and lower abdominal region. They seem to have a preference for the thickest of human hairs, such as those found on the genital region. However, they can also occur on other suitably haired areas of the body, including the axilla, sometimes amongst facial hair, and even on the eyelashes.

In men they can be seen on the abdomen and armpits, where there is often a thicker covering of hair than is the case for (most) women. They can also on occasion occur in the eyebrows. This is known medically as phtiriasis palpebrarun. However, even if occurring here they probably originate from contact of a sexual nature.

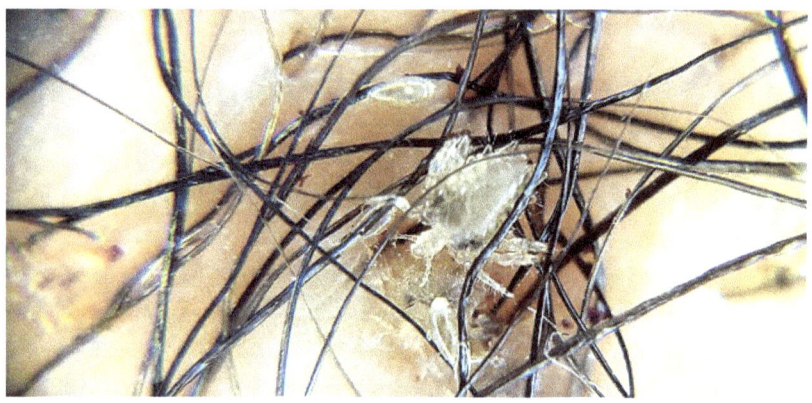

Live louse in pubic hair.

Thus a rather neat form of habitat differentiation appears to have occurred with the three species of human louse. The Body Louse, Head Louse and Pubic Louse have effectively divided the human body into particular segments of parasitism, with each species being found on a separate part.

Pubic Lice are intimately associated with the human body and can not survive for any length of time away from us. Their diet is human blood. Feeding occurs typically five times per day with blood being extracted from hair follicles. The lice bury their heads deeply in the follicles before extracting blood.

The lifespan of the adult lice is approximately 30 days. The complete lifecycle takes one to three months. Adults mate daily, and often over extended time periods. The adult females lay three eggs per day, with as many as 150 to 200 being laid in total over an entire lifetime. These are placed at the base of hairs, often in clusters. In shape the eggs are oval. If you look closely you can see there is a distinct operculum or opening at the upper end. However, more common are to see the empty 'nit' egg cases.

Eggs hatch into larvae after six to ten days. There are three nymphal instars. Pubic Lice are known as what is hemimetabolus insects; this means there is no pupal stage between the nymphal and adult stages. Instead, nymphs simply molt into adults. Maturity is reached within ten days. Larvae and nymphs appear superficially similar to adults, simply smaller versions. Molting occurs successively until they are sexually mature adults.

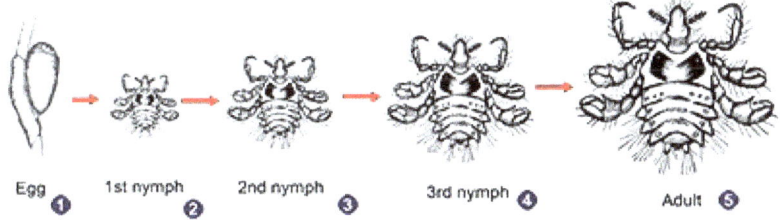

The lifecycle of the Pubic Louse
from egg (1) to adult (5).

Transmission

Generally Pubic Lice are considered a sexually transmitted parasite. Close and intimate contact is required for them to move between people. At the risk of sounding contradictory, despite being the most sedentary of parasites and reluctant to move off the host, they nevertheless readily move from one host to another should the opportunity arise. Therefore transmission of lice by those engaging in sexual activities and suffering infestation is extremely likely due to the dispersive ability of lice. Over 90% of the sexual partners of those infested become infested themselves.

Most commonly transmission occurs because contact is sexual in nature, but not exclusively so. Intimate contact of a non sexual nature can also result in transmission. Transmission is usually of adult lice, rather than those in developing instar stages.

Children are known to sometimes become infested. This is possible through normal close physical contact. This can occur through a child sleeping in the parental bed, and in close proximity to their parent.

It can also occur through close physical play; children lack awareness of personal space and especially in the young close physical contact can result in their spread. However, given the generally sexual nature of the transmission, if found in the genital regions of children this may possibly be an indication of child abuse. The presence of these parasites should be noted by those with safeguarding duties.

A particularly widespread urban myth is that infestation through use of toilet seats is possible. Although indeed technically possible, this is extremely unlikely given how attached to the human host these lice are. You would have to be extremely unlucky and unfortunate to become infested in such a way.

Symptoms

The thought of being parasitized is rather unpleasant; so the idea of having parasites present in the most intimate and sensitive of our anatomical locations is particularly revolting to most people.

The most common symptom experienced by those with Pubic Lice is intense itching around the genital region where they are found. This is often accompanied by reddening. This is often not due to the actual parasitism and feeding of the lice themselves, but rather due to a hypersensitivity reaction caused by lice presence and feeding. The lice inject saliva into hosts prior to feeding which may trigger this reaction. The body reacts to defend itself, and the reaction causes the itching. This sensation is typically worse at night, probably due to increased warmth in the affected area due to bedding making the itching more noticeable.

However, symptoms are not felt on initial infestation, typically developing two to four weeks later. However, in cases of re-infestation reddening may develop within a few days.

A distinctive symptom of Pubic Lice infestation is known as maculae ceruleae; the skin around lice bites takes on a characteristic blue-grey coloration. Additionally small red macules may appear across the area of infestation. They are most commonly seen on the abdomen and thighs.

Maculae ceruleae result from the feeding action of lice; saliva injected into hosts contains enzymes which convert bilirubin into biliverdin creating distinct blue markings. Scratching may result in the development of lesions which are liable to secondary bacterial infection.

Those who are sexually active and who have multiple sexual partners are at most risk of becoming infested with Public Lice. Use of a condom does not reduce risk of transmission of Pubic Lice.

73

Urban myths

Perhaps inevitably because of the predominately sexual nature of how Pubic Lice are transmitted, and the desire to self-treat when infested, a variety of myths have developed around how to eliminate them. The sexual nature of most cases means there may be a reluctance to seek medical advice or treatment and a greater willingness to consider alternative and unconventional treatments.

Although some of these remain popular, they are probably of limited effectiveness. Shaving of the genital region is maybe an obvious solution for those affected. However, this will not eliminate these lice. It does however make life more difficult for them.

Another popular old wives tale is that hot baths are a good remedy. However, again this does not kill any human lice, and may result in severe scalding. Another commonly recommended self-administered treatment is to use either Listerine or vinegar; there is no clinical evidence that these are effective. They are not to be recommended.

Maybe one of the reason so many urban myths have developed surrounding Pubic Lice is because they are relatively rare. Studies consistently show that only between one and five percent of the population are infested. Typically, about one percent of those sexually active harbor Pubic Lice.

In an interesting study Armstrong and Wilson (2006) reported that the popularity in genital shaving may have reduced rates of incidence. This could well have helped, hindering the Pubic Louse in its movement and in completing its lifecycle. It could however be merely an interesting correlation.

SELECTED REFERENCES

General

Leone PA. Scabies and Pediculosis Pubis: An update of treatment regimens and general review. *Clinical Infectious Diseases*. 2007;44:S153-S159.

Salavastru CM, Chosidow O, Janier M, Tiplica GS. European guideline for the management of pediculosis pubis. *Journal of the European Academy of Dermatology and Venereology*. 2017;31(9):1425-1428.

Evolution

Reed DL, Light JE, Allen JM, Kirchman JJ. Pair of lice lost or parasites regained: the evolutionary history of anthropoid primate lice. *BMC Biology*. 2007;5(1):7.

The Brazilian and the Pubic Louse

Armstrong NR, Wilson JD. Did the "Brazilian" kill the pubic louse? *Sexually Transmitted Infections*. 2006;82(3):265-266.

The Sarcoptic Mite

Rashes and itchiness can be caused by a number of conditions. The medical term to describe intense itching is 'pruritis'. But this somewhat odd and cold term does not do justice to the agony we have all experienced when we have an area which just won't stop itching, and when you have that horrible sensation that scratching just seems to make it worse.

There are many possible causes of intense pruritis, one of which is scabies. This is a skin condition caused by infestation with a parasite; the Sarcoptic or Scabies Mite. The alternative name for scabies, 'the itch', describes the principle symptom experienced by those suffering this condition. The intense itching is accompanied with a characteristic rash. However, it is not actually the mite which is the principle cause of these symptoms. Instead the symptoms are caused predominately through the immune response which is triggered by the presence of mites. The body's act of fighting against the invader results in inflammation and soreness.

It took science a long time to realize that scabies was caused by a parasitic mite. This is because Sarcoptic Mites are minute in size. They can hardly be seen with the naked eye. It took the development of the microscopic lens for them to become readily apparent. However, the characteristic rashes these parasites caused were identified much earlier, even if the reason for them was unclear.

Pictures depicting scabies rashes are thought to have been produced by the Ancient Egyptians. Passages in the Bible are thought to maybe refer to treatment of scabies. For example in the King James Bible Leviticus 13 and 14 refer to bathing to alleviate skin diseases such as leprosy, and maybe scabies.

Despite the small size of the mite it appears that they were possibly known about before the development of magnifying lenses. It is thought that Aristotle might have identified them. In his work he describes 'lice of the flesh' which would escape from pimples that were burst; this sounds suspiciously like Sarcoptic Mites. A more likely bet for the first actual observation were those of Avenzoar, (1094–1162) (Ibn Zuhr in Arabic), who describes the mite and crucially made the link to scabies. He thus probably preceded European thinking by several centuries.

Further progress had to await the development of the microscope in the mid 17th century. The mites were first drawn in 1657. Giovan Cosimo Bonomo (1663-1696) and Diacinto Cestoni (1637-1718) are credited with making the link between the mite and scabies disease. These two renaissance scientists were able to examine sailors who were frequently affected by scabies because of the frequent female liaisons they made on their regular shore visits. They observed lice and eggs on afflicted sailors and made the link with the skin condition.

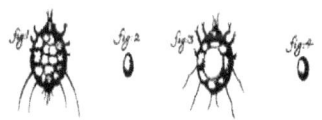

**The first pictures of the
Sarcoptic Mite made by Bonomo.**

Further progress was made in Paris by the research group led by the French dermatologist Jean Louis Marc Alibert. One student was able to make further drawings of the mites. Another, Simon François Renucci, came from Corsica where he had noted that Corsican peasants extracted mites using a fine needle. He was able to demonstrate this technique in Paris, thus providing medical scientists with a method to extract parasites for easy study for the first time.

He also developed what is known as the 'ink burrow test'. Sarcoptic Mites burrow under the skin causing burrows which typically run some centimeters. The French student realized that by placing a washable ink over the affected area, then washing it away, the burrows become filled with ink and thus visible. This provided an easy and reliable method for medical diagnosis. It continues to be used in some parts of the world today.

Taxonomy

The Sarcoptic Mites Latin name is *Sarcoptes scabiei* var. *humanis*. This is an Arachnid, classified within the Subclass Acari. Humans are not the only species to suffer from this parasite. However, each variety is species specific. Thus, the human version is found only on humans.

Possibly most well known to most people are the canine variety. In dogs they are the cause of Sarcoptic mange, with affected dogs losing copious amounts of fur and having distinct bald patches mainly around the rear.

Although transfer of the canine variety to humans is possible, such infestation is self limiting in nature with the parasites not succeeding in establishing themselves on human hosts.

Physical appearance

In appearance Sarcoptic Mites appear rounded and globular. The body is broadly oval shaped and flattened dorso-ventrally. Mites are often described as being 'tortoise like' in appearance.

Sarcoptic Mites are barely visible to the naked eye, females measure 300 to 500 micrometers; males being considerably smaller. There are four leg pairs; legs I and II end in sucker like appendages. Those of leg pairs III and IV of females, and III of males, end in hair like seta.

Lifecycle

These mites feed on skin detritus. The mites complete the entire lifecycle on humans, and they are unable to survive for any length of time away from us.

Adult females burrow into the layer of skin just below the surface, where they lay eggs. These eggs hatch into larvae after about 50 to 53 hours. The larvae undergo a series of molts, before attainment of adult size.

Each life stage takes approximately three days and thus it takes approximately ten to 14 days for them to reach maturity. The lifecycle is repeated every two weeks.

Actually the number of parasites on most people affected by them is relatively small, with most having an average parasitic load of only ten to 15 mites. However, these are prodigious parasites and are capable of rapidly building up in numbers in a short period of time. Thus a small number of unlucky people can harbor many hundreds of the mites.

Sarcoptes scabiei mite.

Symptoms

As already described the classical symptoms of scabies are the development of rashes and the presence of intense pruritus.

Generally symptoms occur in two phases. The initial symptoms are caused by the actual presence of the mites and their action of burrowing into the skin. This tends to be rather mild and somewhat localized. These symptoms develop within a few days of initial infestation as the females begin to burrow into the skin. Areas most commonly affected include the wrists and the areas the mites are actually present. During this stage the actual number of mites is rather small. Rashes are localized and limited. These symptoms typically subside within a few days of initial infestation.

However, some weeks later much more severe symptoms occur. These are not so much the result of the mites themselves, but due to our immunological responses which kick in to the presence of the mites. Much more widespread rashing occurs, typically about four to six weeks following the initial infestation.

These rashes occur in a characteristic arc running from the groin, under the arms and on the breasts. They are sometimes referred to as the 'circle of Hebra' following the Austrian German Dermatologist who described them. Skin rashing can extend from the finger webs, wrists, underarms and lower abdomen, axillary folds and genitalia and lower abdomen. Females may suffer around the breasts. Both sexes may experience rashing around the genitals. These rashes can be accompanied with painful and itchy lesions. The rashes and itching does tend to die down as the body becomes sensitized to the mites. However, the mites remain present and the person can continue to spread them onwards.

The phrase 'the seven year itch', used to describe disillusionment with marriage, is thought to have its roots from scabies infestation. The phrase is conventionally thought to refer to the feeling of wanting a change after a number of years with the same partner. The initial excitement of marriage has passed and partners begin to wish to seek excitement and change. Thus they could become infested with scabies and experience the characteristic symptoms of intense itching.

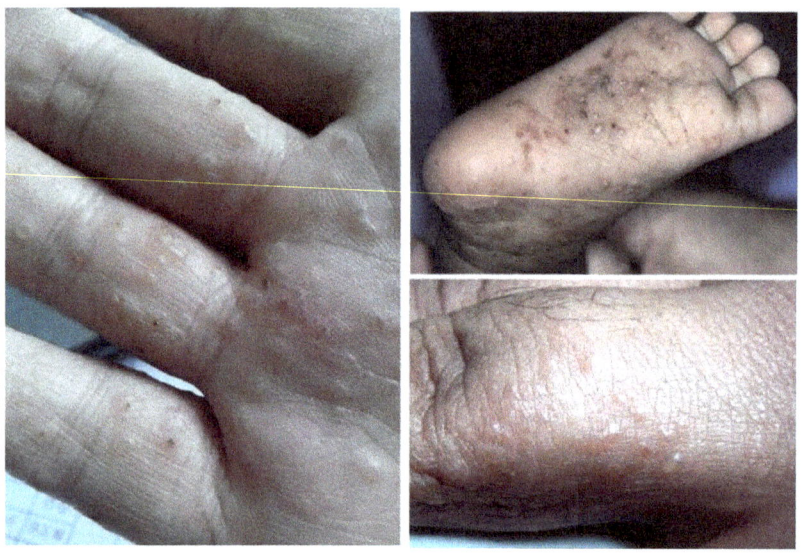

Above: Commonly affected parts of the body include hands, wrist, armpits, groin, and in children the feet.

Scabies is sometimes classed as a sexually transmitted disease. It was notorious as a condition in seamen and soldiers, who had ample opportunity to engage in illicit behavior. However, confusion as to the extent that sexual activity is required to contract it was unclear, as in some cases it was clear that this was not necessary. During the Second World War the British authorities were concerned as to the potential loss of fighting men scabies could cause and funded research into its transmission.

The British Entomologist Kenneth Mellanby was engaged and established a scabies research center in Sheffield, England. A large mansion like Victorian home present in a well to do suburb close to the university was allocated. His task was to study how these mites were being transmitted, and whether such transmission was really sexual in nature or not. But how could you study how the mites were being transmitted?

Mellanby engaged conscious objectors who wanted to help the war effort but who did not want to engage in actual fighting. Today such experiments would be considered morally dubious. However they were not as extreme as the experiments being carried out in the name of medical science by the Nazi regime by a long way.

A number of experiments were conducted. For example, volunteers would use beds in which previous scabies patients had slept in. Other volunteers were intentionally infested with mites so that the course of the infestation could be studied and recorded in detail. Mellanby even speculated on suggesting adultery to volunteers in order to study the mite; a step which was later not taken.

The results of the experiment showed that sexual contact was not necessary for scabies transmission. It was possible to contract mites simply through coming into close contact with bedding or clothing that contained mites.

This requirement for close contact is underlined by the fact that one group most susceptible to scabies today are nurses. These often come into close physical contact with patients. When nurses are infested the mites are commonly found around the wrists. Nurses who handle patients, often hold them around the wrists and this wrist to wrist transmission is common.

Scabies is also common in barrack situations, or where people comes into close contact with each other. Barracks provide the ideal opportunities for them to spread. Thus also often afflicted are the homeless. Being unable to change bedding and remaining in the same clothing for extended periods of time means they are particularly vulnerable to these parasites.

Despite Mellanby's work there still continues to be stigma related to scabies, with many still associating it with physical contact of a dubious kind. This is not helped by patients often being referred to sexual health clinics; much would be said for it being treated as a routine skin condition instead.

Various treatments have been tried throughout history, however many have attempted to alleviate the symptoms of the rashes and itchiness rather than combat the mites themselves. Sulfur has formed the basis of many treatments, it soothes the skin and relieves the itchiness. Benzyl benzoate has a long history of use for scabies and other skin conditions.

In the developed West scabies is considered a minor problem. There might be sporadic outbreaks centered on old people's homes, for example, but few cases crop up. However, in other parts of the world it is a major problem. Scabies is endemic to large parts of the globe, particularly in tropical areas where climatic conditions favor the parasite.

The World Health Organization estimates that at any one time more than 200 million people are suffering the effects of scabies infestation. It is particularly a problem in children. Adults typically build up some resistance to the parasites, having experienced repeated exposure while young. Thus although they may continue to harbor them, they do not typically suffer. They have effectively grown used to scabies. But in children this has not occurred and symptoms can be severe, leading to illness.

The effect on child well-being should not be underestimated. Recurrent infestation is common. Prevalence levels of between ten and 70 percent are reported in some contexts. In 2017 the World Health Organization listed scabies as one of the neglected tropical disease, emphasizing the importance of this disease.

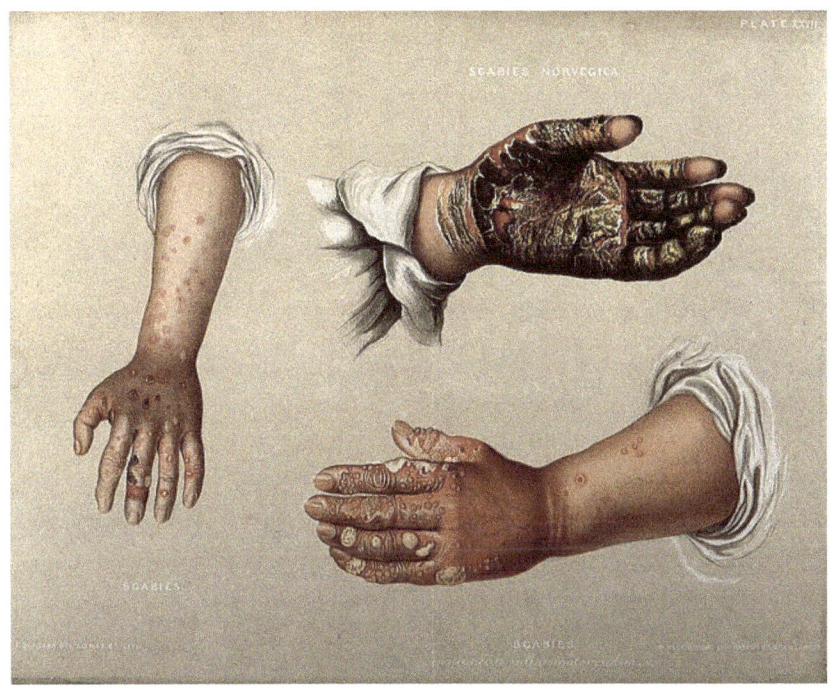

Illustration of hand with scabies.

This is especially so for the Aboriginal populations of Australia. The climate of this part of the world is particularly favorable to the parasite. The way of life of these traditional Australian communities means extended families often live in the same dwelling and in extremely close contact to each other. This facilitates transmission of Sarcoptic Mites. The problem is often worse in children and can have a substantial detrimental effect on education.

SELECTED REFERENCES

General reviews
Chandler DJ, Fuller LC. A review of scabies: an infestation more than skin deep. *Dermatology*. 2019;235(2):79-90.

Chosidow O. Scabies. *New England Journal of Medicine*. 2006;354(16):1718-27.

Heukelbach J, Feldmeier H. Scabies. *The Lancet*. 2006;367(9524):1767-74.

Historical account
Currier RW, Walton SF, Currie BJ. Scabies in animals and humans: history, evolutionary perspectives, and modern clinical management. *Annals of the New York Academy of Sciences*. 2011;1230(1):E50-60.

Modern prevalence in developing world
Romani L, Steer AC, Whitfeld MJ, Kaldor JM. Prevalence of scabies and impetigo worldwide: a systematic review. *The Lancet Infectious Diseases*. 2015;15(8):960-7.

BLOOD FEEDERS

Parasites that live elsewhere, but utilize our blood.

9

Human Fleas

We maybe associate fleas in humans with old dirty tramps and homeless people. However, most fleas which are found on humans are there simply through accident. The most commonly seen fleas on humans are those that specialize on dogs and cats. This is understandable given the often close association we have with these animals.

Taxonomy

Fleas are small wingless insects All fleas are parasitic requiring a host from which they can take blood meals. They live on the exterior of hosts, and survive by drawing blood. They are placed within their own order, the Siphonaptera. There are approximately 2,500 species which are classified within 16 families. They parasitize a range of mammals and birds. However, only a few species are found in association with humans.

The House Flea (*Pulex irritans*) is often considered as the 'human flea'. But despite the name, it can parasitize a wide range of hosts, not just humans. Actually more commonly found on humans are those associated with our pets. The Cat Flea (*Ctenocephalides felis*), can actually be found on both cats and dogs. It is often found in our homes. Similarly, the Dog Flea (*Ctenocephalides canis*) can also be found on dogs, cats and us!

Physical appearance

Fleas are small in size, approximately at most 0.5 millimeters. This means they are easily missed and overlooked. They have a flattened profile which allows them to easily move between fur or feathers. The external cuticle is tough; this offers protection and aids retention of water. In color fleas are reddish-brown. The body is covered with short brisk hairs, which easily get caught up in host fur, further aiding retention. These are backwards pointing, meaning the flea can still move forwards along host hair.

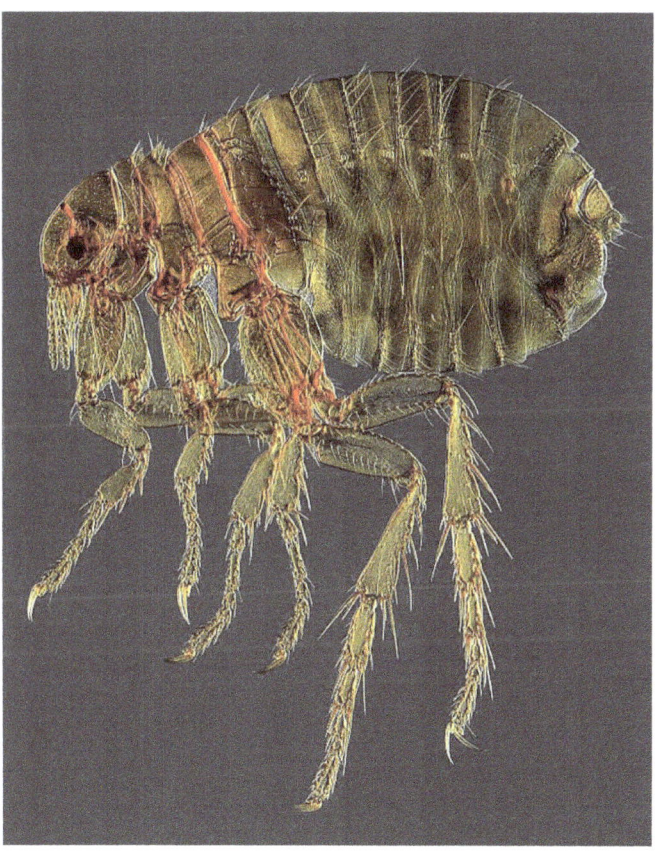

Previous page: A high resolution image of a female *Pulex irritans.*

Lifecycle

Fleas are obligate parasites; they can't live without a host! They have a four stage lifecycle. Eggs are white in coloration and can be laid either on the host or in the habitat in which the flea occurs. These hatch into small larvae, which have no legs and are worm like. These eat bits of debris and organic matter, which may come from the host such as dead skin, but might simply be material present in the environment.

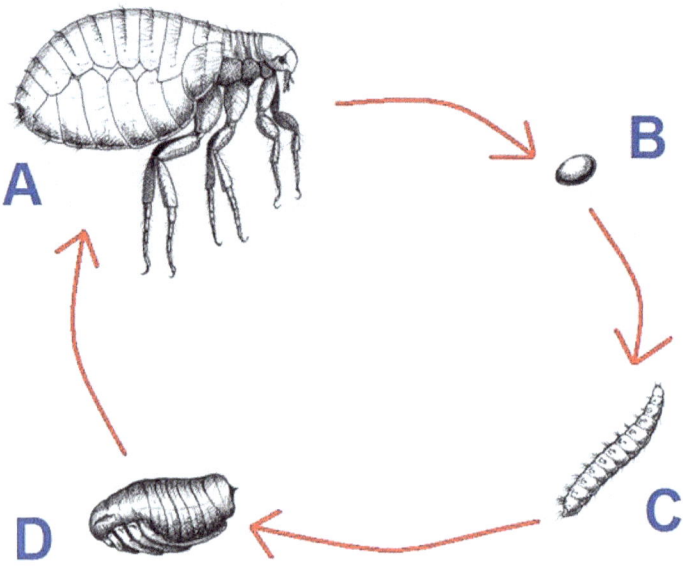

Human flea (*Pulex irritans*) lifecycle: a) Adult. b) Egg. c) Larva. d) Pupa.

Larvae undergo typically three molts, then become pupae. These hatch into the adult form. The length of the lifecycle depends on the environmental conditions at the time. It is typically about ten to 14 days in total.

Adults can typically live for two to four weeks. Adults need to feed off a host. They detect potential hosts using carbon dioxide, heat, and movement as cues. They mate on the host.

Fleas are renowned for their jumping ability! The hind legs are well developed and larger than the frontal pairs. The legs contain a special protein which can be compressed, then suddenly released providing an almost explosive burst of energy. This provides fleas with the ability to jump 150 times their own body length. Thus despite being wingless, they have no trouble moving around. This jumping ability is of obvious utility in helping fleas move onto new hosts.

Sign and symptoms of fleas in humans
Humans are not the preferred host of Cat and Dog Fleas. They might remain on us for some time, and even take a blood meal, but soon depart. Nevertheless the bites can cause itchiness and skin irritation. Often small red raised bumps occur at the bite site; these can be sore and itch. On occasion, and particularly in susceptible individuals the area around the bite can become swollen and inflamed.

Most commonly affected parts of the body are the feet, ankles and legs. This is because these are the parts of the body fleas can most easily reach. They can be difficult to diagnose though, because the symptoms are characteristic of so many parasites.

Fleas as vectors of disease

Human fleas can be vectors for a range of conditions. These include plague (*Yersinia pestsis*), but also a range of illnesses caused by Bartonella bacteria and also a Rickettsial illness known as flea-borne spotted rickettsia.

Plague has literally plagued humankind throughout our history. There is evidence that Neolithic hunters suffered it. Famous was the Justinian plague that occurred in the Roman Empire during the sixth century, hastening the empires decline. Also well known were the plague epidemics of the 14th century, known as the 'Black Death'. Plague epidemics continued in Europe right into the 18th century.

But what exactly caused plague remained a mystery. There was much speculation that it was the result of bad air.

It took until 1894 for the causative agent and vector to be discovered. In this year a plague epidemic struck Hong Kong. Frenchman Alexandre Yersin isolated the bacteria that caused plague. A year later Paul-Louis Simond who was working in India, discovered that fleas harbored plague, and transmitted it to mammals.

Fleas in pets

A survey by Bond et al. (2007) found that the prevalence of fleas was greater in cats (21%) than in dogs (6.8%). Just under a third of cats which were infested showed symptoms, and about half of dogs. Interestingly half of the pet owners where fleas were present were unaware of the infestation.

The extent of the flea problem varies year to year. Population abundance is weather and climate dependent. Mild winters and warm humid summers promote the survival of fleas over the winter and encourage breeding in the summer. Abundances can peak in autumn following a summer suitable for breeding.

The obvious sign that a pet has fleas is constant scratching. There can also be hair loss in localized areas, as well as skin lesions and red inflamed skin. It is recommended to groom your pet regularly and examine it carefully while you do so to spot signs of fleas. Special pet combs with fine teeth can be bought to aid in this.

If a problem is spotted then the first step is to treat the pet, and a variety of insecticidal treatments exist. But this is not enough alone. Fleas are not only present on the pet, but also earlier stages of their lifecycle are present in the general environment. Cleanliness helps, with vacuuming of carpets and furniture helping remove these parasites. Special chemical treatments can also be purchased for use on upholstery around the house. In affected pet households populations can quickly build up, particularly in warm autumn months.

Do parasites help each other? Flea larvae will eat tapeworm eggs if they have the opportunity, and actually pass tapeworms onto animal hosts later. Therefore fleas might not be the only problem a pet owner faces!

SELECTED REFERENCES

Bitam I, Dittmar K, Parola P, Whiting MF, Raoult D. Fleas and flea-borne diseases. *International Journal of Infectious Diseases*. 2010;14(8):e667-76.

Bond R, Riddle A, Mottram L, Beugnet F, Stevenson R. Survey of flea infestation in dogs and cats in the United Kingdom during 2005. *Veterinary Record*. 2007;160(15):503-6.

Royal Society for the Prevention of Cruelty to Animals (RSPCE). How to get rid of fleas. 2024. Available from: www.rspca.org.uk/adviceandwelfare/pets/general/fleas

Zietz BP, Dunkelberg H. The history of the plague and the research on the causative agent *Yersinia pestis*. *International Journal of Hygiene and Environmental Health*. 2004;207(2):165-78.

10

Mosquitoes

Mosquitoes carry some of the deadliest diseases in terms of deaths annually. According to data from the World Health Organization in the World Malaria Report, in 2022 there were 249 million cases of malaria globally. There were over 600,000 deaths. This occurs pretty much annually. Yet malaria rarely makes the headlines. These deaths go mostly unnoticed by the media and the general world.

But malaria is only one of the conditions that mosquitoes are the vectors of. There are others which if not quite as deadly as malaria, still affect a large number of people each year. These include filariasis, chickungunya, dengue fever, yellow fever and the Zika virus.

This makes mosquitoes arguably our most important parasite. This chapter provides a generalized overview of mosquito biology; please be aware that the details for each specific species may vary. The idea is to convey a general idea of mosquito biology.

Taxonomy
Mosquitoes are Dipterous insects. Mosquito taxonomy can be somewhat complicated, and as for other classifications is sometimes revised according to taxonomic fashion.

Generally, the mosquitoes are considered as those species within the Culicidae family. This contains 3,200 to 3,600 species, depending on how you count and classify them. Generally three subfamilies are recognized, the Anophelinae, the Culicinae, and Toxorhynchitinae. Each subfamily is divided into various tribes.

Selection of some of the most important species of Mosquito

Species	Important diseases transmitted
Anopheles gambiae	Notorious malaria vector.
Anopheles arabiensis	Malaria
Culex piepiens	Lymphatic filariasis West Nile virus
Culex quinquefaciatus	West Nile virus St. Louis encephalitis (SLE)
Aedes aegypti	Dengue fever Chikungunya fever Zika fever Yellow fever
Aedes albopictus	West Nile virus

The most important species in terms of disease transmission are *Anopheles gambiae* and *Anopheles arabiensis*, both of these are important vectors of malaria and lymphatic filariasis. Also important in malaria transmission are *Culex piepiens* and *Culex quinquefaciatus*. And not to be forgotten are *Aedes aegypti* and *Aedes albopictus*. Although these are the most important species, a great number of mosquito species can act as vectors of disease. It must be pointed out that although these are called 'species', in many cases there are actually species complexes, containing several very similar species.

Physical appearance

In appearance adult mosquitoes have long slender bodies and long dangly thin legs. There are three distinct body regions; the head, thorax, and abdomen. The head contains sensory structures; there are two large compound eyes, a pair of antenna, two maxillary palpi, and a proboscis.

The thorax is subdivided into three sections. Each section has paired legs. The middle section also contains the wings, while the final section contains halteres; these are structures which provide stability in flight. The abdomen is made up of ten sections and contains the reproductive and excretory organs.

In contrast the larvae have no exoskeleton, instead the tissue is soft and pliable. However, the differentiation into body parts is still clear. However, there are hardened plates in the head section making these tough and hard.

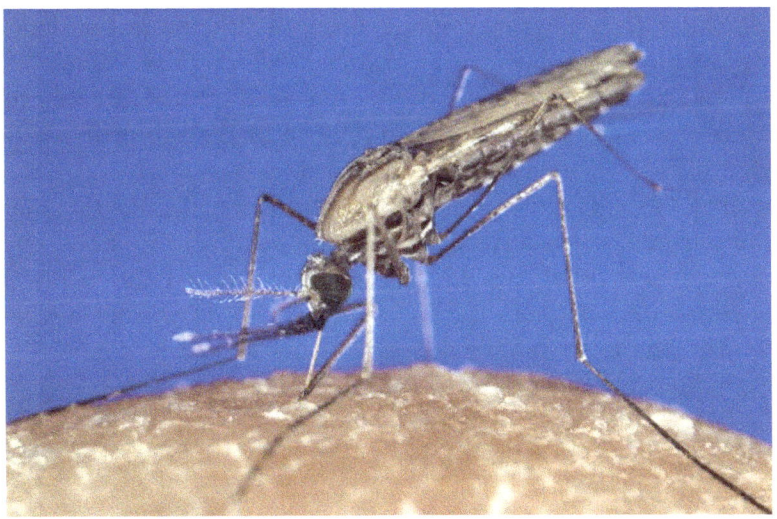

Anopheles gambiae **mosquito feeding. One of the most important malaria vectors.**

Lifecycle

Males have a shorter lifespan than the females, living only about a week, whereas the females live for over a month. The mosquito lifecycle can be divided into four parts; egg, larva, pupa, and adult.

Females lay between 30 and 300 eggs, which they do on vegetation amongst standing water. Sometimes eggs are laid together as 'rafts' which float on the water surface. The requirement for water is an essential part of the mosquito lifecycle.

Within 48 hours however the eggs hatch out into larvae. Essentially larvae are the growing stage of the mosquito. Larvae swim in water and eat organic material. Sometimes they will eat live matter and even each other. They undergo a series of molts, a total of four, growing bigger with each one. The larvae are distinctive; they have long tubes which reach to the water surface and allow them to breathe.

The next stage is the pupal stage, during which the mosquitoes do not grow any more, but mature and rest. Although the larvae grow, the pupae are simply resting and getting ready to emerge as adults. They can still move though, which they do by flicking their tails.

The final stage is the adult one. Adults emerge from the pupae, coming to the water surface and resting in order to dry out. When the wings have dried they fly off and begin to feed and reproduce. The entire lifecycle from egg to adult takes approximately a month. However, the exact length depends on temperature, humidity and the conditions of the water the mosquitoes live in.

Reproduction

Once the mosquitoes are adult they fly and then mate. However, females require a blood meal for the eggs to grow. They identify potential hosts through odor, heat and emitted carbon dioxide.

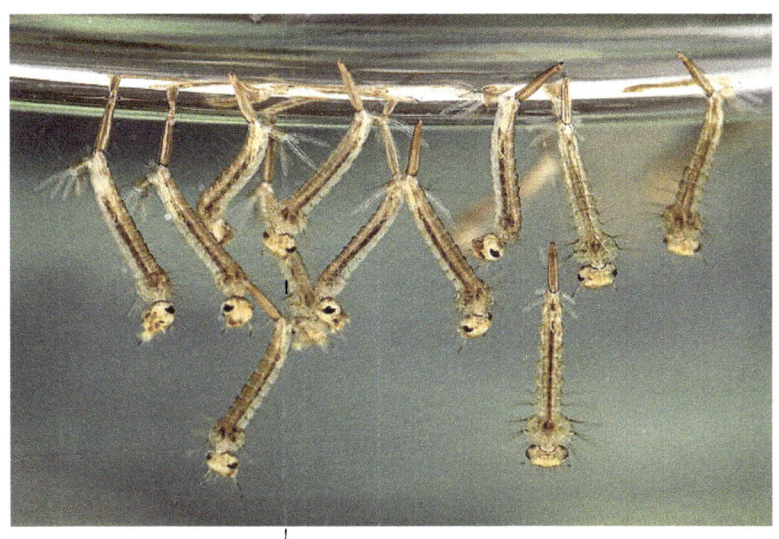

Culex mosquito larvae found in standing water in an Atlanta, Georgia residential area.

How do mosquitoes bite?

Once they have found a potential host to draw blood from, female mosquitoes land on a suitable area. But where is the best place to look for blood vessels? They can sense lactic acids and sugars present on the body surface to indicate where blood vessels could lie.

They bathe the area in saliva, which contains agents to numb it, meaning the host will not sense their feeding, then they insert the proboscis into the skin. The proboscis is needle like and very sharp and can easily penetrate the upper skin. However, many attempts might need to be made until the female finds a blood vessel from which she can remove blood. When she does she injects saliva into the area; this contains anti-coagulants which prevent the blood from clotting, meaning that the blood continues to flow.

The head of the mosquito contains a special pump like structure, which effectively sucks blood from the blood vessel. The female mosquito is a very efficient feeder and capable of removing two to three times her body weight at a single successful feeding.

DISEASES

An entire book could be written, and many have, on the principal diseases transmitted by mosquitoes. Malaria is certainly the most important.

How many people suffer from Mosquito transmitted diseases?

Disease	Number Affected
Malaria	200 million plus cases annually. 600,000 deaths each year.
Dengue fever	96 million develop clinical symptoms annually.
Lymphatic filariasis	Circa 50 million people are infected in total.
Yellow Fever	200,000 annual cases
West Nile virus	Unknown. Likely under reported.
Chickungunya	50,000 to 300,000 cases annually.

Malaria

The causative agent of malaria are single celled parasites belonging to the Plasmodium genus. Despite only being single celled, these are complex organisms. Over 200 different types exist. The lifecycle is complex involving time spent in insect and animal hosts.

Essentially an infected mosquito injects sporozoites into a host. These travel to the liver, where they mature into schizonts which replicate into many merozoites. These invade red blood corpuscles, destroying them and multiplying. A mosquito picks up some of the merozoites, these develop in the mosquito into the sexual form, known as gametocytes. These become fertilized, then develop into oocysts, which contain many sporozoites. The oocysts travel to the mosquito salivary glands, where the sporozoites then get passed on once again

The most important form of Plasmodium is *Plasmodium falciparum* which is widespread in Africa and causes severe disease. Elsewhere *Plasmodium vivax* is the greatest threat, and can lead to recurrent illness.

Malaria is widespread in Africa and South and Central America. The characteristic symptom is fever, accompanied with sweats and chills. There are a range of associated symptoms including headaches, tiredness, and muscle and joint pain. These symptoms develop one to two weeks after the mosquito bite.

However, these relatively mild symptoms can develop into more serious ones, ultimately resulting in death. This is particularly the case in those weakened, such as children or the elderly. Yet treatment, if given in the early stages is usually effective and is relatively simple.

Global Malaria Atlas. Proportion of children two to ten years of age showing, in a given year, detectable *Plasmodium falciparum* parasites (Annually, 2000 to 2022).

However, relatively simple prevention methods have been shown to be effective at reducing the risk of contracting malaria. These include ensuring bed netting is present, as mosquitoes normally bite at night. Window and door screens can stop mosquitoes entering living quarters. Ensuring drainage of wetland areas close to where people live is also an important measure to reduce incidence.

Chickungunya disease

Chickungunya is a viral disease transmitted by mosquito vectors. Chickungunya was first identified following a epidemic in East Africa in the early 1950's. It has subsequently been found around the world and is now a global problem. Periodic epidemics occur in which millions can be affected. Since 2000 these epidemics appear to be becoming more frequent.

Chickungunya is transmitted by mosquitoes of the *Aedes* genus. Within Africa a variety of species may act as vectors, including *A. africanus* and *A. aegypti*. In Asia and India the principal vectors are *A. aegypti* and *A. albopictus*.

Wild mammals, principally small mammals, act as reservoirs of infection. The virus is then transmitted between these reservoir hosts and humans by mosquitoes. When epidemics occur, the virus is transmitted by mosquitoes directly from human to human.

The name chickungunya is particularly apt; it means 'he who walks bent' in the Makonde language of east Africa, where the condition is endemic. The name excellently describes the agonizing pain experienced by sufferers. Following infection there is an incubation period of three to ten days, followed by the sudden onset of fever which may persist for two weeks. The principal symptoms are severe arthralgia, arthritis and joint swelling.

Other associated symptoms are headaches, back pain, vomiting, fatigue, and nausea. Symptoms vary and can range from mild to severe. Acute symptoms can be severe but in the majority of cases clear up, typically within seven to ten days. Although the majority of sufferers quickly recover, a significant number of patients experience long term stiffness and pain. Although mortality rates directly due to chickungunya are low (around 0.2%), complications can significantly contribute to morbidity.

Within Africa it is endemic, occurring in predominately rural areas. A particular characteristic are periodic outbreaks of chickungunya. Significant rainfall appears to be a potential trigger. There were many outbreaks during the 1970's principally in South East Asia, Vietnam, Pakistan, Laos and Burma. There has been a re-emergence from the 2000's with a number of epidemics occurring, predominately centered on urban locations.

SELECTED REFERENCES

General
World Mosquito Program. Mosquito-borne diseases. 2024. Available at: www.-worldmosquitoprogram.org/en/learn/mosquito-borne-diseases

Malaria
World Health Organization (WHO). Malaria. 2022. Available at: https://www.who.int/news-room/fact-sheets/detail/malaria

Chikungunya fever
Staples JE, Breiman RF, Powers AM. Chikungunya fever: an epidemiological review of a re-emerging infectious disease. *Clinical Infectious Diseases*. 2009;49(6):942-948.

11

Midges

Have you ever had a summer evening sat on the patio ruined by midges which seem to attack as the darkness descends? The name 'midge' does not apply to a single species, instead it is a generic name used to denote any of a number of small winged insects which belong to the order Diptera. They are often confused with mosquitoes, which are larger. Although they are not as big as mosquitoes these are nevertheless important parasites; they are the vector of a number of conditions meaning they are far more than simply a nuisance.

Taxonomy

There are a great number of midge species. They are generally considered as being those Diptera which belong to the Nematocera suborder. This suborder contains families which are considered as houseflies and mosquitoes, but also a number which are considered as being 'midges'.

Many of these midge families contain species that are not parasitic at all, and which do not bite. For example, the Chironominae, look like mosquitoes but lack elongated mouthparts in order to remove blood. Species in this family are often closely associated with animals, but in a non-parasitic manner.

However, other families contain blood feeding species including the Ceratopogonidae, which are known as the 'Biting Midges'. And also the Culicoides family which contain some important vectors of diseases. For example *Culicoides imicola* which is found in Africa, transmits bluetongue virus and African horse sickness. *Culicoides obsoletus* in North America can transmit eastern horse encephalitis virus and West Nile virus. Actually differentiating between species can be difficult, and is mostly based on the structure of the genitalia, something that can only be observed under a magnifying scope.

Physical appearance

Midges are small two winged flies. Typically they are only one to three millimeters in length, however larger types do occur. In appearance they are slender with long fragile looking legs. Noticeable are the long antennae which are many segmented. The wings are translucent.

The outline shows the typical profile of a midge.

Lifecycle

Midges go through a lifecycle comprising egg, larva, pupa, adult. The number of eggs laid depends on the species; but it can vary from between a few dozen, to several hundred. Eggs are laid in moist ground or mud. They require high humidity in order to hatch. The eggs hatch into small larvae. Hatching occurs within 24 hours of eggs being laid.

Larvae live in films of water present in the soil, or in small pools found in depressions. They feed on a range of minute organic matter, bacteria, algae and fungi; basically whatever they can find.

The larvae go through several molts before an adult midge emerges after undergoing a final short two to three day pupal stage. If winter arrives, then larvae in the final stage overwinter and commence life again the following spring.

The speed of reproduction depends on the temperature of the environment. In tropical climates there can be multiple generations each year, with a new generation appearing every two to three weeks. In temperate areas however where temperatures are lower there are typically only two or three generations each year.

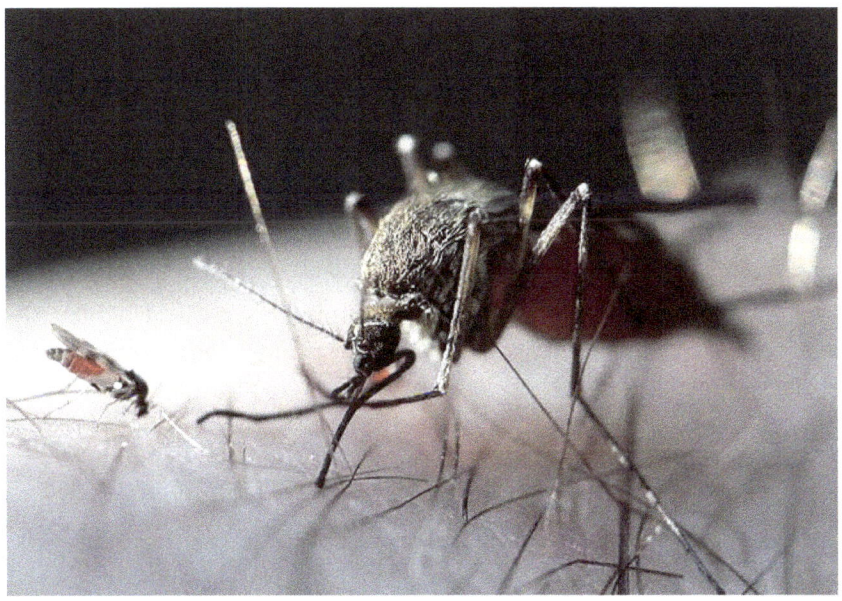

This photo shows the size disparity between midge and mosquito.

Lifestyle

The exact lifestyle varies depending on species. Some midges are active only at dusk or at night, whereas some others are active in the day. The adults feed on nectar in flowers. However, for the female to produce eggs a blood meal is required. Hence they must find a host to parasitize.

Parasitism

A female will feed for three to five minutes, taking only a minute quantity of blood; about two micro-liters. This can cause an allergic reaction in the host, leading to a painful red swelling at the bite point. This can be most itchy causing considerable discomfort.

However, although Biting Midges might be small and the amount they remove as an individual is rather minute, there impact should not be underestimated. Weight of numbers mean they can have a considerable impact. A single bite can cause an allergic reaction, but being bitten multiple times makes things worse.

In some parts of the world, where habitat conditions are favorable, they can cause a considerable problem. For example, in the northern tundra, the landscape is characterized by open plains with numerous areas of boggy marshy ground. This is ideal habitat for midges. Plus, when the early northern spring begins, temperature quickly increase, and there is a paucity of natural predators. This means midge populations can quickly explode.

Large mammals can experience high levels of blood loss, leading to anemia, weight loss, and not rarely death.

DISEASES

Midges act as vectors for a variety of diseases. Notable are those that affect livestock. These include bluetongue virus, which affects mainly sheep and goats. This causes facial swellings with affected animals salivating profusely.

Also of note is African horse sickness which causes head swelling and breathing difficulties in horses. This is highly fatal. Finally Schmallenberg virus affects sheep and cattle, causing abortions and birth defects.

A number of conditions that can be transmitted by midges also affect humans. These include Oropouche fever and filariasis.

Oropouche fever

This viral condition is named after the Oropouche river in Trinidad and Tobago near to where it was first identified in 1955. This is a tropical condition found across South America and into central America.

A variety of wild mammal and birds are natural reserves of the virus. Sloths are sometimes cited as a vector species. Midges, particularly *Culicoides paaensis*, collect the virus through feeding, and can then go on to infect humans.

A particular problem is where humans contract the condition in rural areas, then move to urban ones, where midges quickly spread the virus thus causing epidemics. These epidemics can be large in size. For example, one reported in Brazil led to over 10,000 people becoming infected.

The virus causes fever like symptoms, with those affected having high temperature, muscle and joint pain, which last about a week.

The number of cases is likely to be under reported, but is likely to be in the hundred thousands each year.

Mansonella filariasis

Midges are thought to be one of the vectors of Mansonella filariasis These are parasitic roundworms. Midges transmit these to people while they feed. The roundworms themselves remain close to the insertion site.

However, they quickly start to produce larvae, which are known as microfilariae. These travel around the body causing the host to initiate a host response. Symptoms include headache and fatigue.

This form of filariasis occurs in central Africa and South America. However, much remains unknown about the condition.

Controlling midges

Midges are most abundant when it is warm; just when people are most active outdoors or wish to ventilate their homes. How can midges be controlled?

Netted window and door screens stop entry into houses, and are often used in hot and tropical areas. Also frequently used are insect deterrents such DEET.

Natural products containing lemon or eucalyptus offer an alternative for those who dislike chemicals. Wearing of hats and long trousers stops bites. In areas well known for midges, netted hats can be worn.

SELECTED REFERNCES

Mellor PS, Boorman J, Baylis M. Culicoides biting midges: their role as arbovirus vectors. *Annual Review of Entomology.* 2000;45(1):307-40.

Romero-Alvarez D, Escobar LE. Oropouche fever, an emergent disease from the Americas. *Microbes and Infection.* 2018;20(3):135-46.

12

Reduvid or 'Kissing' bugs

An alternative name used for Reduvid bugs of the Triatoma genus is the 'Kissing' bug; this is because they tend to bite around the head and shoulders, leaving characteristic marks on the face. These parasites are particularly important because they are the vectors of Chagas disease, which is a considerable problem in South America.

Among those to mention and describe these insects was Charles Darwin. There are some who believe that Darwin himself, who in later life was plagued with ill health, may have been infected with Chagas disease and this could have accounted for the chronic ill health he experienced in later life.

Taxonomy
Reduviid Bugs are a variety of different species of the Triatoma, Rhodnius and Panstrongylus genuses. Those of the Triatoma genus are the most common. They appear superficially similar in appearance to Bed Bugs, but are somewhat larger. They are also winged, which means they are much more mobile. Some species have notably striped black and yellow abdomens.

These insects are widespread across the New World and into Oceania. When European settlers arrived they soon became aware of these parasitic insects. However, they were considered more a curiosity than anything.

In 1909 the naturalist Carlos Chagas discovered that it was these insects which were the vectors of disease.

**Rhodnius prolixus, one of the vectors for
Trypanosoma cruzi**

Physical appearance

Like Bed Bugs, Triatoma bugs are true bugs. They measure approximately 1.5 to three centimeters in length. They have a characteristic striped abdomen, which the membranous wings fail to cover when closed. There is a long antennae, differences in which are used in their classification. These insects feed on the blood of vertebrates and have a variety of hosts. They feed at night. The victims rarely realize they are being preyed upon, the insects producing anti-coagulant proteins in their saliva.

In the day when not feeding they inhabit cracks and nooks between stones and amongst walls, only emerging at night.

The feeding by the insects rarely causes any issues. There may be some localized swelling, but often the bites go unnoticed. However, if infected with trypanosomiasis then as the pathogen invades the body, there can be swelling, most typically around the face, with a distinct swelling around the eyelids known as the Romaña sign.

Lifecycle
There are three main life stages for a Triatoma bug; egg, nymph and adult. Adult females lay between ten and 50 eggs, which are placed in nooks and crannies in walls, or even amongst furniture.

Eggs hatch into nymphs. These undergo a number of instar stages, at the end of each they molt into the next one. Nymphs require blood to develop, which they take from mammal and bird hosts.

Finally, they molt into the adult form. Adults continue to take blood meals. The length of the lifecycle can vary depending on the exact species and the conditions at the particular location. It can take several months to complete the egg to adult stage.

Females may mate with a number of males. They are capable of storing male sperm, thus being capable of fertilizing eggs over an extended period of time.

DISEASES

Chagas disease

Kissing Bugs are notorious as the vectors of American trypanosomiasis, or American sleeping sickness. The causative agent is *Trypanosoma cruzi*, a protozoan parasite, transmitted by the bugs as they feed. These Protozoa dwell in the rectum of the insects. As the bugs feed they defecate, releasing infective Protozoa. These are often rubbed into the wounds caused by the biting of the bugs, and then go on to infect the host.

The alternative name for American trypanosomiasis is Chagas disease, in honor of Carlos Chagas who deciphered the condition in Brazil in 1909. Infection is caused by a flagellated Protozoa. Once a person is infected, most commonly through a bite from the Triatonid bug, two phases can occur. First an acute phase may occur in the first weeks or months. Those experiencing this may experience fevers, headaches, aches, vomiting and diarrhea. These symptoms are typical for a number of conditions, thus those infected are often missed as it is mistaken for something else. Some of those infected do not have any symptoms whatsoever.

A relatively common symptom of this stage is the Romaña sign, which is a swelling of the eye. Often symptoms experienced in this first acute phase pass within a few weeks, although a small proportion will experience severe illness.

However, all because illness has subsided does not mean that the Protozoa have gone. They remain present causing damage over the long term. In the next chronic stage of the illness, most sufferers do not have any obvious symptoms. This may persist for decades. However, approximately 20 to 30 percent of those infected develop heart related problems such as heart enlargement, heart failure or a sudden cardiac arrest.

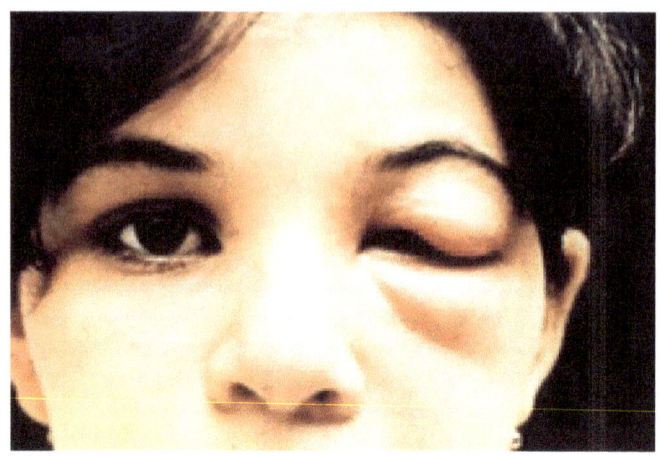

Romaña sign.

Chagas disease is mainly, but not exclusively a disease of rural areas. Dilapidated housing conditions favor the bug, but in modern housing made of concrete there is nowhere for it to reside. The best way to prevent Chagas disease is to improve housing conditions.

SELECTED REFERENCES

Rassi A, de Rezende JM. American trypanosomiasis (Chagas disease). *Infectious Disease Clinics.* 2012;26(2):275-91.

C

OCCASIONAL AND
ACCIDENTAL PARASITES

Parasites that are found on us by mistake or which
do not intend to parasitize us

13

Bots, Blow Fly and Screw Worms

When an animal dies it provides an ideal habitat for a number of parasites. A dead body is essentially a bag full of nutrients, all encased in a protective environment. Such an environment is ideal for the development of insect larvae. The image of something dead being surrounded by flies is almost a stereotype. Bodies can be crawling with maggots.

Although we most commonly associate maggots with when something is dead, parasitism of live hosts is also possible. The medical term for human parasitism by maggots, whilst the host is still alive, is myiasis. 'Mya' is Greek for fly. The maggots come from Dipterous fly parasites, which lay their eggs on hosts which then hatch out into the maggots.

Myiasis is most commonly a problem in agriculture; flies lay eggs in feces which is present upon animal fur or fleece. This provides a substrate from which maggots can emerge. Sheep farmers are well aware of the problem of 'strike', where animals are literally eaten alive by maggots. This is a common problem in spring when warm temperatures and rainfall can be conducive to maggot emergence and survival.

Human cases are mostly centered in tropical areas of the globe. Such locations have the climatic conditions which favor the Dipterous flies which cause the problem. Examples, include the

Human Botfly (*Dermatobia hominis*), which is found in the Americas and which parasitizes human tissue. Another myiasis causing parasite from the Americas is the Screw-worm Fly, (*Cochliomyia hominivorax*). These flies lay eggs in open wound material, from which the larvae hatch out of to feed on neighboring live tissue. The African equivalent to these species is the Tumbu Fly (*Cordylobia anthropophaga*).

Many types of myiasis are recognized, typically they are categorized on the anatomical location being parasitized. So for example Creeping myiasis is caused by larvae which wander about, causing damage over a wide area.

Wound myiasis is fairly self explanatory, occurring when there is some open wound that provides flies an opportunity in which to lay eggs. Cutaneous myiasis refers to cases where the maggots are found under the skin.

Various human body cavities also provide suitable opportunities, and parasitism in the eye orbits, genital tracts and the intestinal openings are also well recognized.

In this chapter we concentrate on the African Tumbu Fly (*Cordylobia anthropophaga*), sometimes known as the Mango Fly, or Skin Maggot Fly, as it provides a good overall example.

Physical appearance
Tumbu flies are small flies of approximately 0.5 to 0.8 centimeters in length. The body is dark colored, with a stripe. The wings are translucent.

The Tumbu Fly has a wide distribution across Africa, being found predominately in the south and east of the continent.

African Tumbu Fly Adult

Lifecycle

Female African Tumbu Flies lay eggs in dirt, soil or in clothing. The eggs are minute and barely visible; meaning they are often missed. These hatch out into larvae in about three days.

When the larvae sense the presence of a host, they begin to burrow into the skin. Each maggot forms a boil from which it feeds on adult skin and gradually matures.

After about three weeks maggots are mature and emerge, then fall off hosts. Then they develop into adult flies which mate and perpetuate the cycle.

Human botfly:
The African Tumbu Fly Larva.

Symptoms

In humans initially there are no symptoms. The initial burrowing of the larvae into the skin occurs with no pain or sensation. Symptoms typically only develop after several days of infestation. There may be a red marking or pimple where the larva has burrowed in. This develops into a more noticeable reddish boil after about a week. Eventually it may even be possible to see larvae moving and wriggling about under the skin!

Once at this stage associated symptoms may develop including intense itchiness, and a fever as the body reacts by starting an immune reaction.

Treatment

Once an infestation is known larvae can be removed manually. Forceps can be used. It may be necessary to squeeze larvae out before grabbing with forceps. Fluid can also be forced into the boil to help eject the maggot. Some say you can suffocate the maggots by covering the boil with petroleum jelly, but removal as quickly as possible is probably better. Removal should be done by a medical professional and with the use of local anesthetics and painkillers. A single patient may have several or multiple locations where maggots are present, so careful examination is required to find all locations where they occur.

Preventative measures include wearing shoes and leg coverings, as often infestation results from contact with contaminated soil. When camping, not sleeping on the ground also helps. Generally good personal hygiene also helps reduce the risk of infestation.

SELECTED REFERENCES

Hall M, Wall R. Myiasis of humans and domestic animals. *Advances in Parasitology.* 1995;35:257-334.

Jelinek T, Nothdurft D, Rieder N, Loescher T. Cutaneous myiasis: review of 13 cases in travelers returning from tropical countries. *International Journal of Dermatology.* 1995;34(9):624-6.

14

Harvest Mites

Harvest Mites are small arachnid parasites which are common in open and countryside areas. In the UK they are known as Harvest or Red Mites, while in the Americas they are more commonly known as Chiggers. Problems are caused by the larval stage, which seeks out an animal host upon which to feed. Humans are merely accidental hosts.

Taxonomy
Harvest Mites are not a single species, instead there are a number of different species belonging to the Trombiculidae family.

Physical appearance
Chiggers are small in size, approximately 0.5 millimeters, and have a bright red or orange color. The larvae are practically microscopic, meaning that they often go unnoticed.

Habitats
These parasites are common in forests and woodland. They are most abundant in early spring when temperatures begin to rise, and vegetation is at its most rush and rich.

Lifecycle
The lifecycle comprises egg, larva, pupa and adult stages. Only the larvae are parasitic.

Adults overwinter and on their emergence in spring the females lay eggs. These hatch into immature larvae after about a week. These initial larvae later mature into feeding larvae which attach to mainly mammal hosts and feed off skin tissue. They have special enzymes to break down skin and release cells for easy feeding. The larvae can be found within the general environment.

Symptoms
Human become parasitized accidentally. The bites can cause painful red itchy marks. This is not that uncommon in those who go camping, or enjoy early spring sunbathing for example.

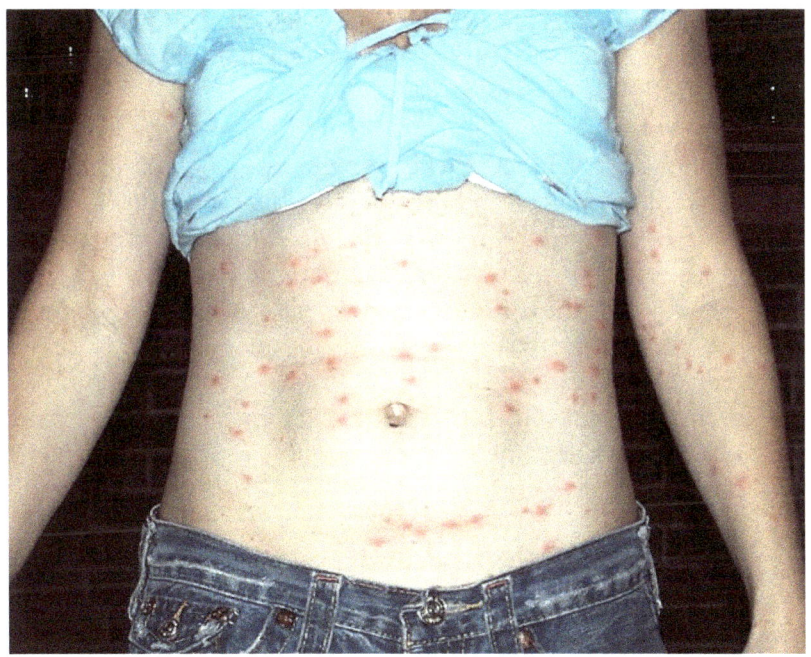

Above: The distinctive bite marks of 'chiggers'.

15

Ticks

The 1990 film Arachnophobia was billed as a 'comic thriller'. The plot involves a small Californian town which was invaded by aggressive large deadly spiders accidentally imported from Venezuela. The film was a hit. But in real life we don't have to worry about deadly spiders, do we?

Technically ticks are blood sucking arachnids, so could legitimately be considered as 'spiders' and thus be possible candidates for such a movie. However, any movie director would probably be unimpressed with them; ticks are rather small in size which would make them appear most nonthreatening on screen. This diminutive size is however deceptive; these parasites can carry serious diseases which can have a real impact on the people who become infected with them. In fact in the northern hemisphere the illnesses transmitted by ticks are the most commonly occurring vector borne diseases occurring in humans.

But ticks are not actually parasites of humans. We are merely, for the most part, accidental hosts. The principal hosts are wild animals; small rodents and large mammals such as deer. Some ticks are most opportunistic though, and in Europe the common *Ixodes ricinus* tick parasitizes a wide range of both mammals and birds and reptiles.

This chapter looks at these arachnid parasites, concentrating on *Ixodes ricinus,* which is commonly known as the Common Sheep Tick or Castor Bean Tick, which is found in Europe and into temperate Asia. We also looks at two of the main diseases they transmit; Lyme disease and tick-borne encephalitis.

Taxonomy

Taxonomically ticks are placed in their own order called the Metastigmata, sometimes called Ixodida, which is within the phylum Arachnida. Ticks are somewhat similar to mites, however unlike mites ticks require a blood meal at each stage of their life.

There are two important families of ticks. Those belonging to the family Ixodae are known informally as the 'hard' ticks. There are over 700 species in this group. They possess a hard outer casing, known as a scutum, hence the naming.

The other family are the Argasidae. This is a smaller grouping with their only being about 200 species within it. They lack a hard scutum and thus are soft and leathery; hence the informal name of 'soft' ticks.

The word 'tick', as used to denote a small parasitic arachnid, is derived from the Middle English word 'tike' or 'teke'. This comes in turn from the Old English 'ticca' or 'ticia'. These words are Germanic in origin, where the name used is 'teek' or 'zecke'. The Dutch is not dissimilar being 'tik' or 'tikken'. The Dutch words mean a light touch or pat. In Middle English the word meant a 'light tap'. It is interesting to speculate whether this refers to the sensation caused by a moving tickling tick running across the skin, or instead refers instead to the penetrating bites which these parasites cause.

Physical appearance

The picture below shows a Sheep Tick (*Ixodes ricinus*). In profile ticks have a generally teardrop shape. Basically a tick can be divided into the large body section, the smaller head, and the legs.

The capitulum or 'head' contains the main feeding apparatus. The body can roughly be divided into two main sections. The frontal part of the body is known technically as 'idiosoma' while the rear portion is known as the 'opistoma'. However, unlike in mites these sections are often indiscrete in ticks. This is especially the case in the soft ticks where the head is often indistinct from a dorsal position because it is held underneath the body.

**Adult female *Ixodes ricinus* tick.
About 3 millimeters long.**

There are four pairs of legs, just like in spiders. These come from the frontal portion of body.

In color ticks have a rich brownish red color, but when engorged the abdomens of the females take on a light grey coloration. What can't be seen in this picture is the flattened profile, which is not apparent in most images. Being flat is particularly useful for a blood sucking parasite; it allows them to nestle close to the skin of its host between feathers and fur, thus directly reach the source of blood.

In size adult males may be 2.5 to 3 millimeters long, females are larger being 3 to 4 mm long. However, this is unfed, when engorged females can be over 1 centimeter in size.

Ticks have a shield like structure offering protection to the body on the dorsal (upper) side. In males this covers the entire body, but in females it is only seen in a small part near the front. This is because when females take large blood meals the rest of the body needs to enlarge and this would not be possible if it was encased in a hard covering. Around the side of the body is notable edging known as the lateral grooves. Across the surface are small hairs.

A frequently used feature used to recognize different ticks species by biologists are spurs; distinct sharp like hooks coming from the first segment of legs. These probably aid attachment to the hosts. Biologists distinguish between internal and external spurs depending where they point.

Auricular, which are analogous to 'eye's, are also present. On the ventral 'belly' males have a plates offering protection around the genitals and body.

The mouthparts are known as the capitulum. As you might expect these are designed for feeding off blood. The hypostome is a long needle like structure used to puncture hosts and pump in salivary fluid then remove blood. It is serrated along its edge. Another feature of the head parts are the palps, which are types of 'feelers' helping manoeuvre the hypostome. The palps are long. There are also a pair of cutting cherlicerae.

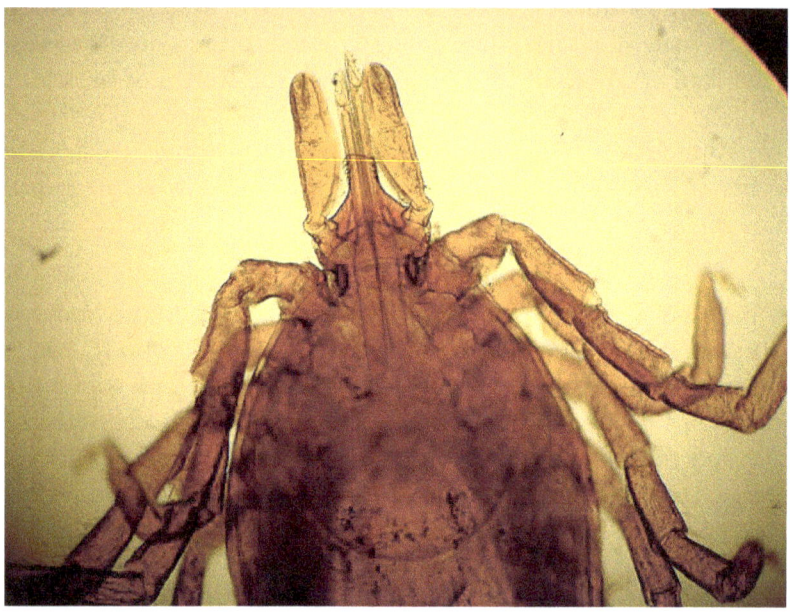

Above: Tick mouthparts. Here you can see the palps on either side of the hypostome.

Overleaf: Here you can see the tick hypostome with serrated edge.

Lifecycle

A tick lifecycle can be divided into four stages; eggs, larvae, nymphs, and adults. *Ixodes ricinus* is a three host tick; meaning a new host and round of feeding is required to progress from larva to nymph, from nymph to adult, and at adult stage before eggs are laid. The lifecycle requires three years to complete.

Eggs: Females lay several thousand eggs over a month. Following egg laying females die. They are laid in moist vegetation. Often they are laid in large clumps.

Larva: The first life stage is as a larva. These are somewhat tick like in appearance, but only have six legs and not the adult complement of eight. In size they are small, not much larger than a grain of sand. Larvae feed on mostly small mammals such as rodents, living in woodland areas.

Nymphs: Larvae molt into nymphs, which are still immature but larger than larva. They are however, still quite small in size. These undergo another bout of feeding. Humans can be parasitized by these and because of their small size this often goes unnoticed.

Nymphs are effectively small adults, and have four pairs of legs, they are sexually immature thus lack genital opening and are smaller than adults being typically 1.4 millimeters long.

Adults: Adults preferentially parasitize large mammals such as deer. Once on the host males mate with females which are also on the host and which are feeding. Unlike females, males may feed a number of times from a number of hosts. They use pheromones to detect females. Mating may take over a week. Following mating females feed for 14 days then drop off the host to lay eggs..

Adult females feed, often enlarging greatly in size. There are many images on the internet of enlarged females feeding upon animals. They take on a typically light grey appearance, the abdomen being engorged with blood.

Females feed to such a degree because they are about to undergo egg laying and require the resources to do so. Once they have finished their substantial meal they drop off into vegetation and begin to lay eggs. They can be most fecund. They are reported to lay up to 100 eggs. The eggs then hatch into larvae and the whole process starts again.

The whole Sheep Tick lifecycle is somewhat protracted. This is because of the number of stages the ticks must go through. Successfully finding a host can take a substantial length of time. Ticks have to get lucky three times. Thus the whole process can take a long time. If lucky ticks can complete their whole lifecycle in as quickly as 3 years, if unlucky it can take 7 years.

Questing

Obviously to obtain a blood meal Sheep Ticks must seek out a host, and it must do this three times during its lifetime. This is an endeavor not guaranteed with success.

However ticks do act to increase their chances. The behavior of finding a host is known as 'questing'. Ticks will climb to the top of vegetation, for example the tips of grass or the edges of a bush and simply wait. The front legs are exposed outwards to help it sense passing hosts. When one does pass by the tick will quickly try to attach.

The front legs possess special sensory cells known as Haller's Organ. These sensory cells help detect carbon dioxide, heat and possible ammonia, all indicative of a passing host. Spreading the forelegs helps expose these cells to these substances.

Questing in *Ixodes ricinus*.

The problem with questing is that it leads to ticks drying out. Ticks are extremely sensitive to desiccation. Being exposed on open vegetation does not help them retain moisture. Thus they will typically only 'quest' when conditions are moist enough. In warm and sunny weather they descend deeper into the vegetation allowing them to re-moisten and await more suitable climatic conditions. Should conditions prove unconducive, or a host fail to pass before a tick has

133

replenished its food resources it will die. However, there are thousands of ticks so the chances are that some will find a host.

Feeding

Ticks take feeding very seriously. Adult females can feed for over a week. Once on the host they will seek out those parts of the body most conducive to drawing blood from the host without being detected and removed.

In mammals, such as deer, ticks often feed off the head and around the ears. In adult humans they often crawl to the groin, or in women the breasts. In children they favour the head and neck.

Once they reach the preferred location they get ready to feed. The palps are used to sense the best place where to insert the needle like hypostome. The hypostome itself is serrated meaning once inserted it is difficult to remove. The hypostome is inserted deeply within the host integument.

How exactly is the hypostome inserted? Firstly the cherlicerae run over the surface integument of the host. There rubbing creates a small depression in the host integument which eases later insertion of the hypostome. They flex and extend repeatedly. This stage of initial attachment can take up to an hour.

In this initial phase of attachment a sticky substance is emitted around the mouthparts which solidifies and forms a kind of 'cement' firmly attaching the tick to the host. This firmly attaches the tick to the host, to such an extent that removal is hardly possible except with the most vigorous of forces. Argasid ticks do not generally produce this cement like substance; but feeding is much quicker in these ticks so such attachment is probably not needed.

Next the chelicerae contract in a breaststroke manner, which drives the hypostome firmly into the host. Each stroke ratchets the hypostome more deeply into the host. Backwards pointing teeth on the hypostome prevent removal.

Ticks are known as pool feeders. The capillaries of the host are cut, causing a loss of blood in the surrounding host tissue, thus a pool of blood is formed. It is this that ticks feed upon. Within the head of ticks there are complicated salivary glands and pump like structures. Initially salivary material is pumped into the host. This contains anticoagulants and painkiller, which aid retrieval of blood and mean the host will not feel the tick. Chemicals are produced to combat the host immune system.

Sadly, it is within this salivary fluid that those bacterial pathogens which cause Babesia and Lyme borreliosis are found, and these are inserted into the host at this stage. The tick does not 'intend' to do this, it simply occurs as they begin to feed. Transmission of these bacteria is of no benefit to the tick.

Feeding is a protracted business; but then ticks only do so a few times in their lifetime, so it pays to get it right. Adult females will feed over several days once attached. Blood removal itself will not start until the tick is well attached. Typically the tick is attached for over 12 hours before salivary material starts to be pumped into the host and before blood is removed. The amount of blood extracted can be substantial but this is not all ingested. Ticks filter blood removing unwanted water and ions, returning what is not needed in the saliva back to the host. Thus although ticks can become greatly engorged, they do not ingest unnecessary amounts of water, only the most sustaining parts making up blood.

Argasid ticks have repeated feeding bouts interrupted by periods of egg laying. However, Ixodid ticks have much longer term feeding, but typically only one feeding session per life stage.

135

Feeding can be divided into two distinct phases. Initially there is a slow phase lasting up to seven days during which ticks increase in weight ten fold. This is followed by a much more rapid absorption of blood with a further ten fold increase in weight over only ten hours.

Engorged tick.

SUMMARY OF MAIN
MOUTHPART STRUCTURES

The head contains the main structures associated with feeding. Both hard and soft ticks have these mouthparts, however, they are most obvious in the hard ticks. They are best seen from the ventral side of the ticks. The mouthparts can be divided into several discrete structures;

- **Chelicerae:** The name 'chelae' is Latin for claw, which is actually a misnomer, as these structures are used for cutting rather than as a claw to grab. The ticks uses these for initial attachment.

- **Palps:** Around the outside of the mouthparts are two sensory structures, essentially 'feelers', which help maneuver the mouthparts and aid in sensory detection. These are called the palps.

- **Hypostome:** Notable in ticks is the hypostome, which is used for extraction of blood. This is essentially a tube used for sucking of blood. It is situated more ventrally than the other structures of the mouthparts. Often they are barbed with sets of spikes along the length; in the same manner in which a screw has a threading meaning it will not come lose once embedded, so these barbs help ensure that the hypostome will remain firmly in place once feeding is initiated. The teeth are known as 'rutella'. The hypostome is essentially a needle like tube; ideal for blood removal.

Sheep Tick Counterparts

Ixodes ricinus is the most widespread and often commonest in Europe, and into temperate Asia, but it is not the only tick species in this area. There are many others. Also of interest is the Marsh Tick *Dermacentor reticulatus*. A number of tick species which specialize to a greater extent upon particular hosts also occur, such as the Hedgehog Tick *Ixodes hexagonus*.

In the USA the counterpart of the Sheep Tick is the Deer tick *Ixodes scapularis*. This occurs across the eastern half of North America. The principal hosts are White-tailed Deer. This species is the most important vector of Lyme borreliosis in the America.

DISEASES

Lyme disease

In the mid 1970's an unusual cluster of cases of children with mysterious juvenile rheumatoid arthritis were reported from the town of Old Lyme, Connecticut. Medical professionals identified this as a potentially new disease, but little was known about what caused it. The new conditions was given the name of the place it had first been identified; Lyme disease.

It was only several years later, in 1982, that microbiologist Willy Burgdorfer working at the Rocky Mountain Laboratories, National Institutes of Health (NIH), in the USA, identified that the causative agent was a bacteria, which was later named after him as *Borrelia burgdorferi*.

People have probably been experiencing Lyme disease for centuries, if not thousands of years. Its identification as a specific condition was hindered by the various manifestations of the disease and its often latent nature. Symptoms can vary widely between individuals. Those affected often experience an untypical 'summer cold'. The most characteristic sign of infection is a characteristic skin rash; called an Erythema migrans, or more commonly a 'bulls eye' rash because of its characteristic shape.

However this does not always occur, and when it does is often missed. After a short period of usually mild illness the condition disappears. But the bacteria remain, and can cause longer term serious health problems in later years. These can include heart complaints, neurological issues, and in severe cases death.

Deciphering the nature of the condition was difficult. Initially the features were unknown and the effectiveness of treatment uncertain. The situation was not helped by the fact that actually there is more than one type of bacteria causing Lyme disease. Different subtypes exist. Although generally similar, each subtype has somewhat differing characteristics. Tests for one subtype failed to identify others. This led to confusion, with tests suggesting no infection, yet patients still experiencing illness. The latent nature of infection, with illness occurring recurrently, also raised problems. Some symptoms, such as tiredness and malaise, are fairly nondescript and the result of a range of conditions, so pinpointing when Lyme disease is present, or some other condition can be problematic. These factors have led to Lyme disease being treat with some skepticism amongst the medical profession.

Although Lyme disease occurs across temperate Europe and North America the causative bacterial agent of disease often differs between these locations. In the United States the main bacterial strain present is *Borrelia burgdorfi*. However, in Europe, three variations occur; *afzelli*, *garelli*, and *burgdorfi*. This means symptoms vary between the continents and even local locations.

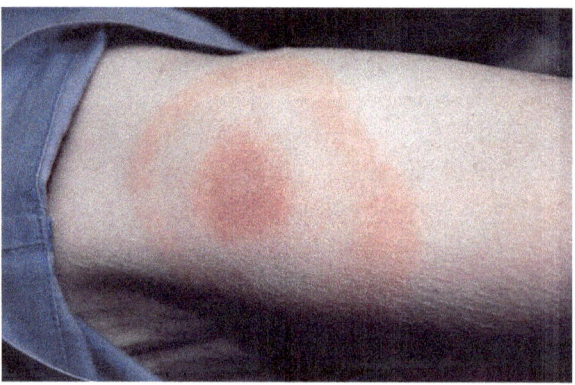

**The characteristic 'EM' or 'Bulls eye rash' of Lyme disease.
But it is not always present, or as distinct.**

140

The 'bulls eye rash' develops at the site of the initial tick bite typically within 10 days of the initial bite. This is an expanding circle of swollen reddish skin centered upon the initial bite site which can reach a diameter of five centimeters. The rash can remain for up to ten weeks. Its presence is often used as a diagnostic feature. However this is not necessarily present; it is seen in only a third of cases.

Tick-borne Encephalitis (TBE)

Another tick borne condition of importance is tick-borne Encephalitis, TBE. Unlike Lyme disease, the symptoms of this can quickly become clearly evident, and be most serious. The causative agent is a virus. Although initial symptoms are fairly innocuous, including fevers and headache, latter symptoms can be much more severe and include meningitis. A notable number of cases lead to death.

TBE mainly occurs in central Europe with the majority of cases being found in the Czech Republic, Germany, Latvia and Slovenia.
However, it is spreading rapidly, possibly as a consequence of climate change or habitat modification. Deer are a vital part in the transmission cycle, and after deer populations were decimated in the 19th century they are making a recovery. But as they do so they are potentially spreading tick borne conditions with them.

SELECTED REFERENCES

Basic tick information
European Center for Disease Control. 2024. Tick borne diseases and preventative measures. www.ecdc.europa.eu/en/publications-data/communication-toolk-it-tick-borne-diseases-and-preventive-measures

Biology
Kahl O, Gray JS. The biology of Ixodes ricinus with emphasis on its ecology. *Ticks and Tick-borne Diseases.* 2023;14(2):102114.

Tick feeding
Richter D, Matuschka FR, Spielman A, Mahadevan L. How ticks get under your skin: insertion mechanics of the feeding apparatus of Ixodes ricinus ticks. *Proceedings of the Royal Society B: Biological Sciences.* 2013;280(1773):20131758.

Lyme disease
Mead PS. Epidemiology of Lyme disease. *Infectious Disease Clinics.* 2015;29(2):187-210.

TBE
Bogovic P, Strle F. Tick-borne encephalitis: A review of epidemiology, clinical characteristics, and management. *World Journal of Clinical Cases: WJCC.* 2015;3(5):430.

Picture Credits

I have used copyright free images or those available in the public domain throughout. Those which require a Creative Commons Attribution or recognition of owner are listed below.

P34: House dust mites (*Dermatophagoides pteronyssinus*). Gilles San Martin from Namur, Belgium. Creative Commons Attribution-Share Alike 2.0

P43: Female *Pediculosis humanis* louse. Daniel J. Drew. Creative Commons Zero, Public Domain Dedication

P44: Body Lice live amongst clothing seems. Courtesy of CDC/ Reed & Carnrick Pharmaceuticals. 1976.

P47: Male body louse. CDC/Frank Collins, Ph.D. Photo credit James Gathany.

P51: Poster: Naomi Baumslag, Murderous medicine: Nazi doctors, human experimentation, and typhus, Westport, Conn., 2005, p. 9 (reproduced). Part of the Wellcome Image collection.

P54: Male (top) and female (below) human head louse. *Pediculus humanus capitis*. Gilles San Martin. Creative Commons Attribution-Share Alike 2.0

P55: CDC/Dr. Dennis D. Juranek. unhatched nit of the parasitic head louse

P68: Pubic louse, close-up of claws, SEM. David Gregory & Debbie Marshall. Wellcome Collection. Attribution 4.0 International (CC BY 4.0)

P70: Live louse in pubic hair: Courtesy of 'Davidoc86'. Creative Commons Attribution 4.0

P80: *Sarcoptes scabiei* mange mite. Coutesy of Alan R Walker. Creative Commons Attribution-Share Alike 3.0

P82: Scabies rashing. Bệnh ghẻ ở bàn chân và tay. Amintoosad. Creative Commons Attribution-Share Alike 4.0

www.ingramcontent.com/pod-product-compliance
Lightning Source LLC
Chambersburg PA
CBHW071718170526
45165CB00005B/2061